METHODS IN MOLECULAR BIOLOGY

For further volumes:
http://www.springer.com/series/7651

For over 35 years, biological scientists have come to rely on the research protocols and methodologies in the critically acclaimed *Methods in Molecular Biology* series. The series was the first to introduce the step-by-step protocols approach that has become the standard in all biomedical protocol publishing. Each protocol is provided in readily-reproducible step-by-step fashion, opening with an introductory overview, a list of the materials and reagents needed to complete the experiment, and followed by a detailed procedure that is supported with a helpful notes section offering tips and tricks of the trade as well as troubleshooting advice. These hallmark features were introduced by series editor Dr. John Walker and constitute the key ingredient in each and every volume of the *Methods in Molecular Biology* series. Tested and trusted, comprehensive and reliable, all protocols from the series are indexed in PubMed.

Chromosome Analysis

Methods and Protocols

Edited by

Eisuke Gotoh

Division of Diagnostic Imaging, Department of Radiology, Japan Labour Health and Safety Organization, Tokyo Rosai Hospital, Ohta-ku, Tokyo, Japan

Humana Press

Editor
Eisuke Gotoh
Division of Diagnostic Imaging,
Department of Radiology
Japan Labour Health and Safety
Organization, Tokyo Rosai Hospital
Ohta-ku, Tokyo, Japan

ISSN 1064-3745 ISSN 1940-6029 (electronic)
Methods in Molecular Biology
ISBN 978-1-0716-2432-6 ISBN 978-1-0716-2433-3 (eBook)
https://doi.org/10.1007/978-1-0716-2433-3

This Humana imprint is published by the registered company Springer Science+Business Media, LLC, part of Springer Nature.
The registered company address is: 1 New York Plaza, New York, NY 10004, U.S.A.

Preface

Chromosomes or genes, the genetic materials composed of large molecule DNA and ptoreins, are the most fundamental element of living things, as there is no creature which has no chromosomes. Chromosomes also have the highly ordered and sophisticated structure and functions compared to other cellular organelles. Chromosomes have an extraordinarily large molecular weight, with a total of ~3 × 10^9 base pairs of DNA (46 in number in humans), and packaging, histone, and non-histone proteins and other constituents, which distinguish it from other cell organelles. Not only are chromosomes a vector of heredity, but they also show important aspects, for instance: (1) control gene expression sequentially through the life cycle of an organism, (2) manage different processes of gene activation in different tissues—failure of such orchestral regulations leads to anomalies in cells such as illness, cancer, or congenital diseases caused by abnormal fetal development or cell death in the human case; and (3) change cell structure drastically during the cell cycle, in particular at mitosis, leading to mitotic condensed chromosomes. Mitotic chromosomes are highly packaged structures that facilitate the transfer of identical genetic information precisely into two daughter cells. Therefore, many researchers are interested in and fascinated with chromosomal research.

More than 150 years ago, the first finding of chromosomes in cells was reported. Since then several breakthrough findings and chromosomal manipulation techniques have been discovered, which have drastically improved our understanding of the chromosomal nature. For example, establishment of the mitotic chromosomes' preparation technique (i.e., cell culture, arresting the cell in mitosis, hypotonic treatment, and squashing and spreading chromosomes in a single plane) revealed that human chromosomes are 46 in total: 22 pairs of autosomal chromosomes (#1 to #22) and a pair of sex chromosomes (X, Y). The discovery of the double-helix DNA structure confirmed the connection between Mendel's rules of inheritance and the behavior of recombination chromosomes of paternal and maternal origins, and the mechanics of sex determination. This finding also revealed that an abnormal number of chromosomes causes congenital anomalies such as Down syndrome (trisomy of #21), Turner syndrome (46XO), or Klinefelter syndrome (47XXY). The G-banding or R-banding method further revealed the two different sequence-specific regions of euchromatin and heterochromatin. This staining technique also allows us to classify chromosomes more precisely. Morgan's work of making linkage maps of genes certified that the chromosome has a linear form and genes line up lineally along the chromosome. The linkage of genes also showed that linked genes do not follow Mendel's law. To honor Morgan's discovery, genetic linkage distance unit, cM (centi morgan), is named after him.

From classical heredity to molecular heredity, two basic principles exist, genetics and epigenetics. Genetics is the study of hereditary phenotype changes that involve alterations in the genetic material, from chromosomal to DNA base level—for example, aneuploidy such as Down syndrome (trisomy of #21) or several types of cancer. On the contrary, epigenetics is the study of heritage phenotype changes that do not involve changes in DNA sequences.

The typical example of this type of phenomenon is DNA methylation or histone modification, such as X chromosome inactivation or several types of cancer.

As this book is titled *Chromosome Analysis*, essential and fundamental chromosomal manipulation protocols were assembled for the readers so that they will be able to research the chromosomes successfully. The book starts with the protocol for the very basic and fundamental technique of preparing mitotic chromosomes to learn, brush up, or confirm the knowledge, not only for the novice who has just stepped into chromosomal research, but also for every artisan in the field where they face some difficulties in chromosomal manipulation in experiments, because this technique is an indispensable one for every person experimenting in chromosomal study. The following chapters are about protocols or topics related to the analysis of chromosomal aberrations induced by mutagens, that is, mitotic index, classification of abnormal chromosomes, sister chromatid exchanges (SCE), or the Comet assay. Further, the recently developed karyotyping method and fluorescent *in situ* hybridization (FISH) technique for whole chromosomes or telomeres are documented. FISH has allowed us to easily distinguish specific gene regions or individual chromosomes; thus, these techniques are now more popularly used than the classical G- or R-banding method. A new technology of premature chromosome condensation (PCC) for obtaining chromosomes effectively is introduced. Various immunohistochemical approaches, including the chromatin immunoprecipitation (ChIP) method, are also described. Finally, recently developed new chromosomal concepts such as histone modification analysis over low-mappability regions and the epigenetics concept of activation/suppression of genes and the resulting carcinogenesis are discussed.

I and all contributors are now convinced that these excellent chapters will be extremely helpful and definitely lead to success in chromosomal research for every person in the field. This book would not have been possible without the contributions of a truly international team of researchers, who are all experts in the field of chromosomes. I wish to express my deepest thanks to all of them for taking the time to describe the details of these captivating protocols. I, finally, would also like to express my sincere appreciation for Emeritus Professor John M. Walker of the University of Hertfordshire, Series Editor of Methods in Molecular Biology, for giving me an honorable opportunity to edit this finest book.

Under the worldwide pandemic fear of COVID-19

Ohta-ku, Tokyo, Japan ***Eisuke Gotoh***

Contents

Preface ... *v*
Contributors ... *ix*

1 A Basic and Simple Chromosome Preparation Protocol ... 1
Eisuke Gotoh

2 Metaphase Chromosome Preparation and Classification of Chromosomal Aberrations ... 9
Takamitsu A. Kato

3 Mitotic Index Analysis ... 17
Takamitsu A. Kato and Jeremy S. Haskins

4 Preparation of Mitotic Cells for Fluorescence Microscopy ... 27
Yuji Tanno

5 Chemical-Induced Premature Chromosome Condensation Protocol ... 41
Eisuke Gotoh

6 Apoptosis Detection Assays ... 53
Ilham Alshiraihi and Takamitsu A. Kato

7 Alkaline Comet Assay to Detect DNA Damage ... 65
Kade D. Walsh and Takamitsu A. Kato

8 Analysis for Sister Chromatid Exchange ... 73
Takamitsu A. Kato

9 Cytokinesis Blocked Micronuclei Aberration Analysis ... 83
Takamitsu A. Kato

10 DNA Damage Foci on Metaphase Chromosomes ... 93
Takamitsu A. Kato

11 FISH with Whole Chromosome Painting Probes ... 99
Takamitsu A. Kato

12 Telomere Aberration Detection by PNA FISH Probe ... 105
Takamitsu A. Kato

13 Nontraditional Method for Telomere Staining by PNA Probes ... 111
Takamitsu A. Kato

14 Visualizing Active Replication Regions in S-Phase Chromosomes ... 117
Eisuke Gotoh

15 Hi-C Analysis to Identify Genome-Wide Chromatin Structural Aberration in Cancer ... 127
Atsushi Okabe and Atsushi Kaneda

16 CUT&Tag Using "Stress-Free" Con A-Conjugated Dynabeads® ... 141
Yasuhiro Fujiwara and Yuki Okada

17 Live Cell Synthetic Histone Acetylation by Chemical Catalyst 155
Shigehiro A. Kawashima and Motomu Kanai

18 Histone Modification Analysis of Low-Mappability Regions.................. 163
Naoko Yoshizawa-Sugata and Hisao Masai

Index *187*

Contributors

ILHAM ALSHIRAIHI • *Department of Environmental & Radiological Health Sciences, Colorado State University, Fort Collins, CO, USA; Department of Biology, Faculty of Science, University of Tabuk, Tabuk, Saudi Arabia*
YASUHIRO FUJIWARA • *Institute for Quantitative Biosciences, The University of Tokyo, Tokyo, Japan*
EISUKE GOTOH • *Division of Diagnostic Imaging, Department of Radiology, Japan Labour Health and Safety Organization, Tokyo Rosai Hospital, Ohta-ku, Tokyo, Japan*
JEREMY S. HASKINS • *Department of Pediatrics and Human Development & Department of Pharmacology & Toxicology, Michigan State University, East Lansing, MI, USA*
MOTOMU KANAI • *Graduate School of Pharmaceutical Sciences, The University of Tokyo, Tokyo, Japan*
ATSUSHI KANEDA • *Department of Molecular Oncology, Graduate School of Medicine, Chiba University, Chiba, Japan*
TAKAMITSU A. KATO • *Department of Environmental & Radiological Health Sciences, Colorado State University, Fort Collins, CO, USA*
SHIGEHIRO A. KAWASHIMA • *Graduate School of Pharmaceutical Sciences, The University of Tokyo, Tokyo, Japan*
HISAO MASAI • *Department of Genome Medicine, Tokyo Metropolitan Institute of Medical Science, Tokyo, Japan*
ATSUSHI OKABE • *Department of Molecular Oncology, Graduate School of Medicine, Chiba University, Chiba, Japan*
YUKI OKADA • *Institute for Quantitative Biosciences, The University of Tokyo, Tokyo, Japan*
YUJI TANNO • *Bioscience Department, Veritas Corporation, Minato-ku, Tokyo, Japan*
KADE D. WALSH • *Department of Environmental & Radiological Health Sciences, Colorado State University, Fort Collins, CO, USA*
NAOKO YOSHIZAWA-SUGATA • *Research Center for Genome & Medical Sciences, Tokyo Metropolitan Institute of Medical Science, Tokyo, Japan*

Chapter 1

A Basic and Simple Chromosome Preparation Protocol

Eisuke Gotoh

Abstract

Chromosome preparation for chromosome analysis is basic and indispensable for wide area of biology such as genetics, medicine, molecular biology, or other many fields. However, it seems that the many artisans in that fields feel hard to get start chromosome preparation technique, as it is something technically demanding or troublesome, or it needs something secret manner to prepare good quality of chromosome spreads. Actually many technical variations among individual laboratories exist. The aims of this chapter is to describe concise and minimal fundamental protocols specifically focused on human chromosome preparation, as step-by-step guiding with instructing essential points.

Key words Cytogenetic study, Chromosome preparation, Collecting blood, Peripheral blood lymphocytes (PBLs), Separation of lymphocytes, Lymphocytes cell culture, Colcemid block, Mitotic index (MI)

1 Introduction

Many kinds of human tissues, either in vivo or in vitro materials, are utilized for chromosome analyses and cytogenetic studies. The choice of tissue depends on the objects, purposes, medical indications, or many other factors. Peripheral blood is the one of most used specimens for chromosome analysis. Primary culture method of peripheral blood lymphocytes (PBLs) is well established. Blood sample is easily obtained by venipuncture with mostly noninvasive manner. Isolation technique of PBLs from whole blood followed by culture of PBLs is a first step for many other in vitro assays. A series of culturing PBLs, cell harvesting, and preparation of chromosome sample from PBLs contains fundamental and essential steps for chromosome analysis technique. Therefore, skill acquisition for chromosome preparation from PBLs is certainly useful for applicable wide range of chromosome preparation technique. This chapter aims (1) to introduce simple chromosome preparation protocol from collecting venous blood and (2) to guide a principle technique of chromosome preparation, step by step with detailed notes based

Eisuke Gotoh (ed.), *Chromosome Analysis: Methods and Protocols*, Methods in Molecular Biology, vol. 2519,
https://doi.org/10.1007/978-1-0716-2433-3_1,

on our chromosome research experiences [1–5]. For learning more about the deep knowledge of chromosomes and its experimental protocols, the readers are recommended to look for many books about chromosomes or biology; for example, see references [6, 7].

2 Materials

Vacutainer® CPT™ centrifuge tube or other equivalents for isolation of mononuclear cells containing peripheral blood lymphocytes (PBLs) (*see* **Note 1**):

1. RPMI 1640 medium supplemented with 20% of fetal calf serum (FCS) for culture PBLs (*see* **Note 2**).
2. 10 μg/ml PHA-P (phytohemagglutinin-P): Dissolve with distilled water to make given concentration (*see* **Note 3**).
3. Colcemid solution in HBSS (Hank's Balanced Salt Solution): Shipping as 10 μg/ml concentration (*see* **Note 4**).
4. PBS(−): Phosphate-buffered saline without Ca++ or Mg++.
5. 0.075 M KCl: Dissolve 7.45 g KCl (M.W. 74.55) in 100 mL Milli-Q water or RO (reverse osmosis) water to make a 1 M KCl stock solution. Take 7.5 ml of the 1 M stock solution and make it up to 100 mL with Milli-Q water or RO water to obtain a 0.075 M KCl working solution. Usually, it does not need sterilizing.
6. Carnoy's fixative (methanol: glacial acetic acid = 3 part vol.: 1 part vol.): Take 3 part of 100% methanol and 1 part of 100% acetic acid (e.g., 150 mL of methanol and 50 mL of acetic acid to make 200 mL fixative), and mix together well. After mixing, the temperature will be increased according to exothermic reaction.
7. 10 mM phosphate buffer (pH 6.8 at 25 °C): Dissolve 136.09 g KH_2PO_4 with 1 L Milli-Q water to make 1 M stock solution A, and dissolve 174.18 g K_2HPO_4 with 1 L Milli-Q water to make 1 M stock solution B. Mix 50.3 mL solution A with 49.7 mL solution B, and then make it up to 1 L with Milli-Q water to make 10 mM working solution. If desired, adjust the pH before the final mess up to 1 L.
8. Giemsa dye solution: Usually suppled from makers as 50% concentration dye dissolved in 100% methanol. Dilute with 10 mM phosphate buffer (pH 6.8 at 25 °C) to make 5% (4–6%) solution (must freshly prepare just before use).
9. Glass slides and cover slips: Glass slides are soaked in 100% methanol until use. Wipe out methanol with lint-free tissue paper.

3 Methods

1. Peripheral blood sampling.
   1. Thread 21–22G hypodermic needle onto sampling folder for a Vacutainer® tube.
   2. Cleanse the venipuncture site at forearm with alcohol-disinfected cotton.
   3. Apply tourniquet and venipuncture, and collect about 5 ml of whole blood automatically sucked in an evacuated tube (*see* **Note 5**).
   4. Remove the tube from folder, gently invert/upright tube several times to mix anticoagulant additive with blood. At this time, three layers consist from top to bottom—whole blood, polyester gel barrier, and density solution—can be seen.
2. Isolation of peripheral blood lymphocytes.
   1. Set Vacutainer® tube with blood in a swing bucket rotor; centrifuge for 30 min at 1500 to 1800 RFC (relative centrifugal force) at room temperature (*see* **Note 6**).
   2. After centrifuged, from upper to lower, about 7 layers consist plasma, and buffy coat contains mononuclear cells and platelets, viscous density solution, polyester gel barrier, density solution, granulocytes, and red blood cells (*see* **Notes 6**).
   3. Aspirate carefully upper layer of plasma and discard. Then, gently collect the buffy coat above the viscous density layer and transfer into a centrifuge tube.
   4. Dilute the buffy coat with PBS(−), centrifuge at 1500 RFC for 5 min at room temperature, and resuspend in PBS(−). Repeat the steps twice. Finally, PBL cells are suspended in 5 ml of RPMI1640 medium supplemented with 20% FBS and transferred to a culture dish or flask. Usually, culture medium volume is the same as collecting blood volume (i.e., start from 5 ml blood; suspend the PBLs in 5 ml of culture medium).
   5. To induce blast formation of T-cell lymphocytes, final concentration of 0.1 μg/ml of PHA-P is added in the culture medium and then incubated for 72 h until the cells reach mitosis (*see* **Note 7**).
   6. Two hours before cell harvest, add final concentration of 0.05 μg/ml colcemid into a medium for arresting cells in mitosis (colcemid block method) (*see* **Note 8**).

3. Cell harvest and chromosome preparation.
   1. Three days (72 h) after cell culture, PHA-P-stimulated T-cells aggregate together to form a mass like a mulberry. At this time, individual cell becomes larger and brighter than the start of cell culture.
   2. Gently pipet the culture medium to loosen and separate the aggregated cell mass. Confirm that the cells absolutely separate. Transfer the medium to a centrifuge tube (*see* **Note 9, 10**).
   3. Centrifuge at 140 g for 5 min at room temperature, discard supernatant, resuspend the cells in PBS(−), and centrifuge again in the same condition. Repeat twice these steps.
4. Hypotonic treatment and fixation of cells.
   1. Suspend the cells in 5 ml of 37 °C pre-warmed 0.075 M KCl hypotonic solution and incubate at 37 °C for 20 min.
   2. Add same volume of freshly prepared and chilled Carnoy's fixative at XX °C. Gently invert/upright the tube several times, and then centrifuge at 140 g for 5 min at 4 °C. After being centrifuged, swollen cells attached on the bottom with faint opaque white color. Discard hypotonic solution/fixative by simply inverting the tube vertically (*see* **Note 11**).
   3. Pinch the top end of tube with thumb, index, and middle fingers. Wave back and forth promptly to loosen the cells completely (*see* **Note 12**).
   4. Add fixatives, centrifuge, discard fixatives, and loosen the cells as described above. Repeat 3 times these steps to complete fixation (*see* **Note 13**).
   5. Finally, resuspend the fixed cell in appropriate amount volume of fixative to prepare chromosome spread. For example, final suspension volume is about 500 μl if the culture volume is 5 ml. This suspension calls chromosome suspension.
5. Preparation of chromosome spread.
   1. Thread yellow tip on micropipette, take 15 μl of chromosome suspension, drop on a pre-cleaned glass slide, and air-dry. At this point, it is recommended to confirm the quality of preparation by just looking with a microscope with low magnification without staining. If the spread is too sparse (messy), suspension volume is too much (less). If so, centrifuge again, discard supernatant, and resuspend with more proper volume of fixative (*see* **Note 14**).

6. Staining, mounting, and observation.
   1. Finally, chromosome spread is stained with 5% Giemsa dye solution diluted with 1/15 M PBS (pH 6.8) for 5 min, and then washed with distilled water and air-dried (*see* **Note 15**).
   2. Prepare 3 series of Coplin jars filled with 100% of xylene. After air-dried, the slides are immersed in the series of xylene and then mounted with a cover glass use with mounting media. Observe under a light microscope with appropriate magnification, confirm the quality of chromosome spread, and score mitotic index (MI, number of mitotic nuclei/number of total nuclei) (*see* **Note 16**).

4 Notes

1. Ficoll-Paque method is traditionally used for the separation of PBLs from whole blood. However, it is somehow technically demanding than the use of Vacutainer® tubes. For the details for handling Vacutainer® tube, please see the instruction provided by the supplier.
2. Usually, culture medium supplement with 10% FCS is used in many kinds of cells. In case of primary culture of PBLs, 20% supplement FCS is much better for PBL blast formation. Heat inactivation of FCS is not usually necessary. If there are some troubles during culture, try heat inactivation of FCS at 56 °C for 30 min. The use of antibiotics for prevention from bacterial contaminations, as use plastic sterile culture wares and the artisan is skilled in aseptic handling. If the case of bacterial contamination troubles happened, antibiotic solutions such as penicillin-streptomycin solution for cell culture should be used.
3. Two types of phytohemagglutinin, PHA-P and PHA-M, are usually supplied. PHA-P is a more purified type. PHA-M has lower purity than PHA-P and still contains some amount of root nodule bacteria *Rhizobium*. Therefore, the use of PHA-M may cause contamination, and activation of T-lymphocyte is not effective as PHA-P.
4. Different product name for equivalent colcemid solution is provided, for instance, colcemid, demecolcine, or KaryoMAX solution.
5. Instead of using Vacutainer® tube, collect the blood with 10 ml of disposable syringe and 21–22G hypodermic needle. Then, carefully layer the blood onto the Ficoll-Paque solutions followed by centrifugation.
6. Centrifuge must be done at room temperature; otherwise, individual fraction does not separate well. If the centrifuge is chilled at use, turn off cooling and centrifuge in a ballast for a while to raise the temperature.

7. Peripheral blood lymphocytes are usually arrested at G0 stage of cell cycle unless stimulated by antigens or lectins. Lectins stimulate lymphocytes in a nonspecific manner, and lymphocytes enter in cell cycle to reach mitosis. PHA-P/M is a T-cell-specific stimulator; on the other hand, ConA (concanavalin A) is a B-cell-specific stimulator. In our experience, PHA-P gives better and stable results than ConA.
8. Concentration of colcemid affects the degree of condensation, and exposure time of colcemid affects the mitotic index (MI, mitotic cells/total cells ratio). Higher dose and longer exposure time of colcemid may cause increase MI and more condensed chromosomes. Therefore, the adequate combination of dose and exposure time of colcemid should be fixed by experimental design.
9. If the cells are still aggregated and not well separated at this step, the cells keep going through aggregation in the later steps, resulting in bad chromosome spread quality.
10. A round bottom polystyrene tube is preferable. A conical tube is not suitable for chromosome preparation as the cells are too tightly packed to loosen after being centrifuged.
11. After centrifuge following hypotonic treatment, faintly white-colored thin layer of cells loosely adhere to tube bottom. To discard the supernatant, simply invert vertically as quick as possible. Do not slant to discard the supernatant; otherwise, the cell layer will be slip away from the tube.
12. Loose and separate the cells completely. If aggregated cells remain at this step, cells never part in the later steps.
13. At least repeating 3 times the wash/suspension step is recommended. For less than 3 times repetition, spreading of chromosomes quality may not be good.
14. Try half or twice volume of the previous volume.
15. Washing with tap water may result in better fixation of Giemsa dyes than washing with distilled water. It depends on the water quality of local tap water.
16. In case of activation of PBLs from healthy adult donor blood, MI usually reaches around 2%.

References

1. Gotoh E, Yamagami T, Yamamoto H, Okamoto H (1988) Chromosomal assignment of human VIP/PHM-27 gene to 6q26->q27 region by spot blot hybridization and in situ hybridization. Biochem Int 17(3):555–562
2. Gotoh E, Asakawa Y, Kosaka H (1995) Inhibition of protein serine/threonine phophatases directly induces premature chromosome condensation in mammalian somatic cells. Biomed Res 16(1):63–68
3. Gotoh E, Asakawa Y (1996) Detection and evaluation of chromosomal aberrations induced by high doses of gamma-irradiation using immunogold-silver painting of prematurely

condensed chromosomes. Int J Radiat Biol 70(5):517–520. https://doi.org/10.1080/095530096144716

4. Gotoh E, Tanno Y (2005) Simple biodosimetry method for cases of high-dose radiation exposure using the ratio of the longest/shortest length of Giemsa-stained drug-induced prematurely condensed chromosomes (PCC). Int J Radiat Biol 81(5):379–385. https://doi.org/10.1080/09553000500147667
5. Gotoh E (2007) Visualizing the dynamics of chromosome structure formation coupled with DNA replication. Chromosoma 116(5): 453–462. https://doi.org/10.1007/s00412-007-0109-5
6. Human Chromosomes (1995) Principles and techniques, 2nd edn. McGrow-hill, Inc., N Y. https://doi.org/10.1002/mrd.1996.1080430105
7. Clark M, Wall WJ (1996) Chromosomes : the complex code, 1st edn. Chapman & Hall, London

Chapter 2

Metaphase Chromosome Preparation and Classification of Chromosomal Aberrations

Takamitsu A. Kato

Abstract

Chromosomal aberrations are changes in structure and number of chromosomes. Metaphase chromosome can be analyzed by a standard light microscope to detect chromosomal aberrations. Recently, detailed analysis or rapid analysis was possible by using fluorescence probes and fluorescent microscope. The origins of chromosomal aberrations can be errors of DNA repair, cell divisions, and DNA synthesis. Analysis of chromosome aberrations can be used for the wide range of analysis. It includes a basic science connecting DNA damage to cellular death and mutagenesis and diagnostic tools for hereditary diseases and biodosimetry following radiation exposure.

Specific DNA damages produce unique types of chromosomal aberrations. Analysis of chromosomal aberrations enables us to investigate the mechanisms of genotoxic stress. However, one type of DNA damage provides a variety of changes in chromosome structures. It is often confusing. This chapter introduces the standard technique of metaphase chromosome spread preparation and typical classification of chromosomal aberrations.

Key words Chromosomal aberrations, Metaphase spread, Chromosome type aberrations, Chromatid-type aberrations

1 Introduction

Structural change analysis of chromosomes is used to detect DNA damages introduced in interphase cells. DNA damages are introduced exogenously including ionizing radiation, genotoxic agents, and endogenously including oxidative stress, replication errors, misrepair of genotoxic damages, and chromatin condensation. Chromosomal aberrations (CAs), structural changes of chromosomes, are as results of misrejoining or unrepaired DNA damages. These cause abnormal chromosome morphology, that is, visible under microscope. In 1938, Karl Sax published "Chromosome aberrations induced by X-rays" [1]. He observed many types of chromosomal aberrations induced in *Tradescantia* chromosomes after irradiation of X-ray. Hypothetical models of

Eisuke Gotoh (ed.), *Chromosome Analysis: Methods and Protocols*, Methods in Molecular Biology, vol. 2519,
https://doi.org/10.1007/978-1-0716-2433-3_2,

radiation-induced CAs were developed by Lea [2]. Initial cytogenetic studies were developed in plants and insects with squash-fixation-staining methods such as aceto-orcein [3]. Later, mammalian cytogenetics were developed with hypotonic pretreatment [4], alcohol and acetic acid fixation, and air-dry method [5]. This method finally confirmed human total chromosome number as 46 by Tjio and Levan [6]. The development of methods for producing well-spread metaphase figures from cell culture growing mammalian cells has led to wide interest in mammalian cytogenetics.

Initially, ionizing radiation was used for chromosomal aberration analysis. DNA damage and reunion of broken chromosome ends produce chromosomal type aberrations. This model is breakage-reunion hypothesis proposed by Sax and developed by Lea and colleagues [1, 2]. Radiation-induced chromosomal type aberrations can be explained well with this classic theory. Later, exchange hypothesis was proposed to explain chromatid exchanges by Revell [7]. Basically, the initial damage is an unstable lesion rather than a break. Many kinds of CAs lead to mechanical separation problems during cytokinesis and resulting in genetic loss after cell division. As a result, one or both daughter cells die. On the other hand, some kinds of CAs are not lethal at divisions and may be transmitted to next cell generations. Some of these symmetrical aberrations need special staining technique to identify them.

The mechanism of two types formation of CAs, chromosome type and chromatid type aberrations (Figs. 1 and 2), was proposed by Bender and his colleagues [8, 9]. Chromosome type aberrations occur from DNA double-strand breaks (DSBs) introduced before DNA synthesis at S-phase. On the contrary, chromatid type

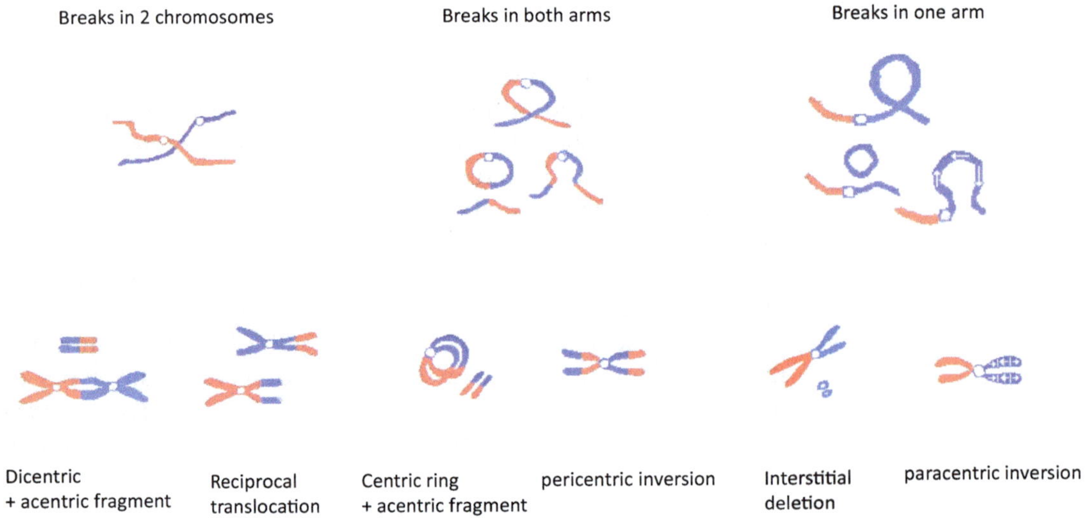

Fig. 1 A simplified model of chromosome type aberration formation

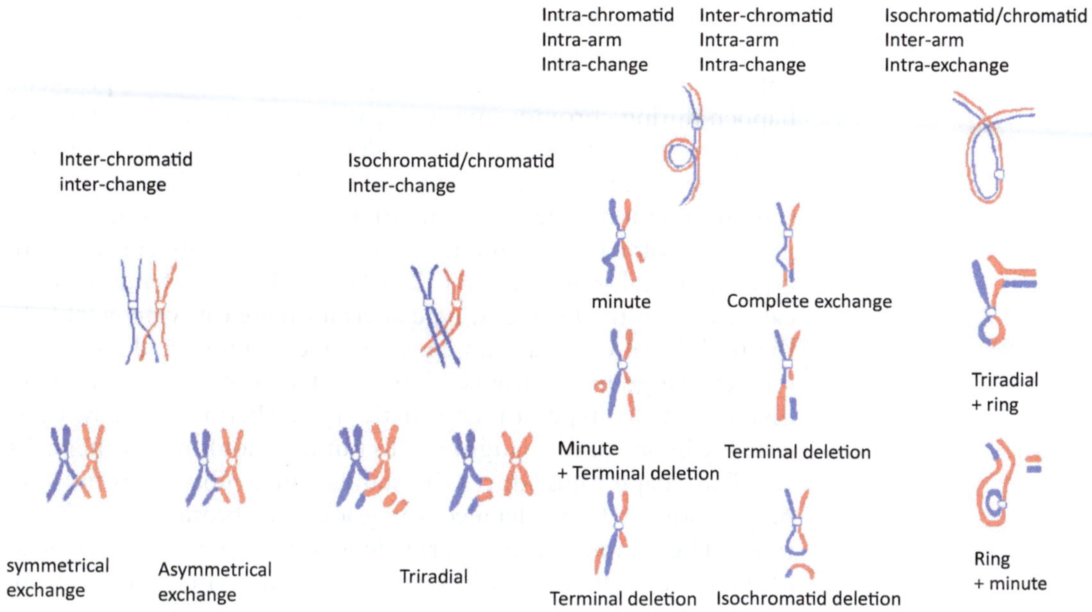

Fig. 2 A simplified model of chromatid type aberration formation

aberrations occur from DNA DSBs introduced after DNA synthesis. Therefore, ionizing radiation (IR) exposure in G1 phase cells produces chromosome type aberrations, and IR exposure in G2 phase cells causes chromatid type aberrations. However, DNA damages could be usually converted to DNA DSBs through DNA synthesis and other cellular mechanisms. UV exposure to G1 cells produces chromatid type aberration by conversion of DNA damage to DSB during S-phase. Chromosome type aberrations are very rare after chemical exposure in G1-phase. Radiomimetic agents like bleomycin, which can directly produce DSB, are exceptions. If metaphase is analyzed at the second or subsequent division after initial chemical exposure, chromosome type aberrations might be observed. Chromatid type aberrations are commonly seen chromosome aberrations after chemical exposure in G1 phase.

Chromosome aberrations cause lethal mechanical problems against chromosome segregation or genetic loss during cell division. The former problems are often associated with two centromeres in one chromosome or tangled during segregation. The examples of these chromosome aberrations are dicentric and centric ring and asymmetrical chromatid exchanges [10]. Cytokinesis failure can be observed as an anaphase bridge. The latter types of chromosome segregation anomalies arise from chromosome without centromere. These chromosome aberrations are acentric fragments such as interstitial deletion and terminal deletion. The formation of acentric fragments can be observed with cytokinesis block-mediated micronuclei formation assay [11]. On the other hand, reciprocal translocation is usually not lethal. Reciprocal translocation may cause de novo gene fusion but does not cause apparent morphological structure problems. Therefore, no problems happen during chromosome segregations or no chromosome loss. Chromosome type reciprocal translocation shows apparently normal appearance, thus not detectable with the traditional Giemsa staining method. Therefore, traditional banding technique or single/multicolor FISH staining is required to identify chromosome type reciprocal translocation [12, 13]. On the other hand, reciprocal exchanges of chromatid type aberration are easily detectable by the traditional Giemsa staining. Even one asymmetrical chromatid type exchange may cause cell lethality. However, in course of cell division, many types of chromatid type aberrations may cause lethality in one of the daughter cells but may not for the other cells.

This chapter introduces the standard metaphase chromosome preparation and the detailed categories of chromosome aberrations. The annotation is to provide a comprehensive information of the kinds of aberrations observable with light microscopy. If the first posttreatment metaphase is analyzed, we can speculate what kinds of damages were produced in the cells at the stage of cell cycle.

2 Materials and Equipment

2.1 Materials

1. Appropriate cell culture media for individual cell such as MEM, alphaMEM, DMEM, and RPMI. Supplemented with fetal bovine serum (FBS) and antibiotic. Prior lot testing is required for FBS. Some lots increase background frequency of SCE formation. Heat inactivation of complement at 56 °C for 20 min is required for FBS.(*see* **Note 1**). Antibiotics such as Antibiotic-Antimycotic (100x) and Penicillin-Streptomycin solution (5000 U/ml) should be added.
2. Phosphate-buffered saline (−) (PBS).
3. Trypsin-EDTA (0.05% or 0.25% trypsin/EDTA-4Na in PBS).
4. Colcemid solution in PBS (10 μg/mL).
5. 75 mM hypotonic potassium chloride (KCl).
6. Carnoy's fixative solution (methanol and acetic acid = 3 vol: 1 vol), prepare at time of use.
7. 5% Giemsa solution. Mix 2.5 ml of Giemsa stain solution with 47.5 ml Gurr buffer solution. Freshly dilute at the time of use.

2.2 Equipment

1. Cell culture facility including biosafety cabinet or clean bench, 37 °C CO_2 controlled incubator, and water bath.
2. Centrifuge, which can rotate 15 ml tubes. No need of temperature control or brake.

3. Aspirator to remove supernatant and trap waste.
4. Upright light microscope.
5. Slide glass, coverslip, mounting medium.

3 Methods

3.1 Metaphase Chromosome Preparation

1. G0/G1 synchronized cells should be prepared on dishes or flasks (*see* **Note 1**).
2. Expose testing agents and trypsinized cells and release cells from G0/G1 phase.
3. Colcemid should be added when cells reach mitosis and wait for another 6 hours (*see* **Note 2**).
4. Collect media with cells and PBS wash and trypsinized cells into 15 ml tube.
5. Centrifuge cell suspension with 200 g for 5 min at room temperature. Discard supernatant by aspiration or decantation.
6. Add 3 ml of 75 mM KCl hypotonic solution and incubate at 37 °C water bath for 20 min.
7. Prepare Carnoy's fixative solution during this incubation time.
8. Add 1 ml of fixative solution and invert tube and mix well gently.
9. Centrifuge cell suspension with 200 g for 5 min. After centrifugation, larger amount of white pellet should be visible.
10. Discard supernatant by aspiration or decantation. Cell pellet should be loosened by tapping tube.
11. Add 1 mL of fixative, and invert tube and mix well; centrifuge and discard supernatant. Cell pellet should be broken by tapping tube.
12. Repeat this process 2 more times.
13. Use yellow pipette tip (100–200 μL); mix well and break pellet gently with pipetting.
14. Take 40 μL of cell suspended solution and drop onto slides.
15. Air-dry for 5 min. Do not disturb slides till dry.

    3.2.1.5. Put slides into freshly prepared Giemsa solution for 5 min.

    3.2.1.6. Rinse slides with running water and air-dry.

    3.2.1.7. After mounting media with coverslip, observe under microscope.

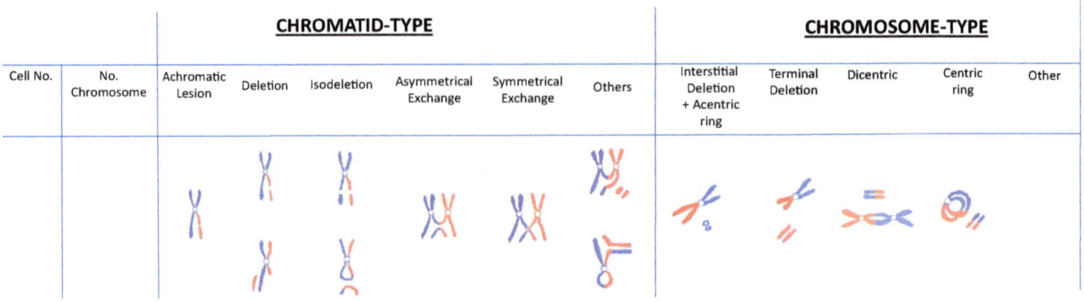

Fig. 3 An example of chromosome aberration categories with representative illustrations. Dicentric or centric ring should be counted with designated acentric fragment

3.2 Analysis of Chromosomal Aberrations

1. Metaphase chromosome can be analyzed under a microscope or after image acquisition.
2. Find well spread and less overlapped metaphase. Count and record chromosome number. Find any chromosomal aberrations. You may focus on specific types of chromosomal aberrations for the research interest.
3. Record observed chromosomal aberrations with categories (Fig. 3) (*see* **Note 3**). At least 50 metaphase cells should be analyzed.

4 Notes

1. This protocol focuses on G0/G1 synchronized cells. Experiments can be performed in the asynchronous cell population or synchronized to other cell cycle phases. In those cases, modifications are required for colcemid timing and repair time after testing agent exposure.
2. Colcemid must be added after enough incubation time, which allows G0/G1 cells reach mitosis. Genotoxic agents often slow down cell cycle time. Therefore, the timing of colcemid treatment can be varied in the samples.
3. The categories of chromosomal aberrations may be varied in the research interests. However, it should be categorized at least as chromatid type break, chromatid exchange, chromosome type break (acentric fragments such as interstitial deletions, terminal deletions), and chromosome type exchanges (centric chromosomes such as dicentrics and centric rings).

References

1. Sax K (1938) Chromosome aberrations induced by X-rays. Genetics 23(5):494–516
2. Lea DE (1946) Actions of radiations on living cells. Cambridge University Press, London
3. McClintock B (1929) A 2n-1 chromosomal chimera in maize. J Heredity 20:218. https://doi.org/10.1093/oxfordjournals.jhered.a103183
4. Slifer EH (1934) Insect development VI. The behavior of grasshopper embryos in anisotonic, balanced salt solutions. J Experi Zool 67(1):137–157. https://doi.org/10.1002/jez.1400670107
5. Hsu TC (1952) Mammalian chromosomes invitro.1. The karyotype of man. J Heredity 43(4):167–172. https://doi.org/10.1093/oxfordjournals.jhered.a106296
6. Tjio JH, Levan A (1956) The chromosome number of man. Hereditas 42(1–2):U1–U6
7. Revell SH (1959) The accurate estimation of chromatid breakage, and its relevance to a new interpretation of chromatid aberrations induced by ionizing radiations. Proc R Soc Lond B Biol Sci 150:563–589. https://doi.org/10.1098/rspb.1959.0043
8. Bender MA, Bedford JS, Mitchell JB (1973) Mechanisms of chromosomal aberration production .2. Aberrations induced by 5-bromodeoxyuridine and visible light. Mutat Res 20(3):403–416
9. Bender MA, Griggs HG, Walker PL (1973) Mechanisms of chromosomal aberration production I. Aberration induction by ultraviolet light. Mutat Res 20(3):387–402. https://doi.org/10.1016/0027-5107(73)90060-2
10. Savage JR (1976) Classification and relationships of induced chromosomal structual changes. J Med Genet 13(2):103–122. https://doi.org/10.1136/jmg.13.2.103
11. Fenech M, Morley AA (1985) Measurement of micronuclei in lymphocytes. Mutat Res 147(1–2):29–36
12. Brown JM, Kovacs MS (1993) Visualization of nonreciprocal chromosome exchanges in irradiated human fibroblasts by fluorescence in situ hybridization. Radiat Res 136(1):71–76
13. Rowley JD (1973) Letter: a new consistent chromosomal abnormality in chronic myelogenous leukaemia identified by quinacrine fluorescence and Giemsa staining. Nature 243(5405):290–293. https://doi.org/10.1038/243290a0

Chapter 3

Mitotic Index Analysis

Takamitsu A. Kato and Jeremy S. Haskins

Abstract

Cellular division is a fundamental process of cellular growth. First, cells replicate their DNA in S phase and then undergo mitosis which, under normal conditions, leads to complete cell division. Moreover, mitotic activity correlates to cellular growth activity. The simplest and classical method to measure mitotic activity (mitotic index (MI)), is the manual counting of mitotic cells among a given cell population of interest. The latter can be accomplished via phase contrast microscope observation. However, Giemsa staining may improve accuracy and consistency. Fluorescence immunostaining targeting specific phosphorylations of proteins at critical cell cycle steps will provide further improved analysis via high-throughput capacity of flow or imaging cytometer. Finally, time lapse image analysis provides quantitative and qualitative metrics delineating the process of cellular division including timing of division, duration of mitosis, and failure to procced through or complete mitosis.

Key words Mitotic index, Histone H3 serine residue 10, Time lapse image analysis

1 Introduction

Cell division is a dynamic event. Deploying phase contrast microscopy allows one to visually observe that attached cells are appropriately detaching and presenting a phenotypically round appearance in an in vitro cell culture system. Typically, contingent upon specific cell line used, within 1 h of observing a round phenotype, a cell divides into two identical daughter cells. Moreover, this division of one cell is commonly referred to as mitosis; daughter cells have identical copies and numbers of chromosomes. Since mitosis is remarkably discernible against interphase under microscope, mitosis is the first identified cell cycle phase that is drastically distinguishable from interphase. Following, the stages of the mitotic cycle were identified as S (DNA synthesis) and G1 and G2 (gaps) by Howard and Pelc [1]. The fraction of cells in mitosis, called mitotic index (MI), was analyzed to determine cellular doubling activity. MI analysis was employed as a simple means to quantify the mitotic "figure" among the tested cell population of interest. Early

Eisuke Gotoh (ed.), *Chromosome Analysis: Methods and Protocols*, Methods in Molecular Biology, vol. 2519,
https://doi.org/10.1007/978-1-0716-2433-3_3,

studies showed the correlation of MI and cellular proliferation rate (i.e., actively dividing, quiescent, or senesced) [2, 3]. Importantly, MI analysis is a powerful cellular technique in that it can be performed on nearly every imaginable cell type or line.

While the prevalence of cancer is increasing in the general population, many of these cancers are malignant and largely heterogenous. Insofar predicting their volatile, malignant behavior has proven tenuous [4]. Although MI analysis does not have the capability to fully characterize the heterogeneity of a disease, it equips healthcare providers a metric to measure the proliferation rate of many cancers. The rate of proliferation proffers healthcare providers prognostic relevance when ascertaining whether a malignancy is highly metastatic or will reside in its local environment. Importantly, MI fraction in breast cancer patient samples was highly correlated to long-term survival outcome [5].

Notably, drugs or chemical agents may increase or decrease MI, depending upon their mechanism of action. Drugs that increase MI such as colchicine were later identified as tubulin inhibitors [6–8]. Moreover, colcemid, a derivative of colchicine, is often used for metaphase spread preparation in cytogenetics [9]. Interestingly, ATR mutants did not change mitotic index after UV exposure due to failure of phosphorylation of CHK1 and activation of G2/M checkpoint [10]. However, most research-related genotoxic agents reduce the MI through reduction of proliferation rate and cell cycle arrest in pre-M-phases. Therefore, MI analysis provides potential insight regarding the cytotoxicity of chemicals and activation of cell cycle checkpoints of interest. Above all, one may conveniently and expeditiously determine the superficial cytotoxicity of a chemotherapeutic or the carcinogenicity of a compound through employing a MI analysis [11, 12].

Mitotic events are highly regulated through a convoluted, networking cascade of kinases. Among them is histone H3, a key constituent in faithful mitosis [13–16]. Specific and imperative phosphorylation occurs at histone H3 serine residue 10 (H3–10) [17]. Histone H3 is phosphorylated at Ser10 or 28 almost exclusively during late G2 and mitosis, promoting the condensation of chromosomes [18]. Detecting phosphorylated histone H3-S10 (pHH3) by immunohistochemistry or immunocytochemistry is useful in the rapid assessment of MI in predicting prognosis of cancer. Distinguishing between apoptotic bodies and mitotic cells may be problematic for less-experienced researchers. However, this is suitably and readily resolved through the pHH3 staining method due to apoptotic cells void of pHH3 signal and mitotic cells presenting pHH3 signal. Phosphorylation at Ser10, Ser28, and Thr11 of histone H3 is tightly correlated with chromosome condensation during both mitosis and meiosis. Recently, a specific antibody was generated against H3 peptide, phosphorylated at Ser10. This phosphorylation can be observable with indirect immunocytochemistry.

Moreover, fluorescent signals can be observed under fluorescent microscope or by using imaging and flow cytometry for high-throughput analysis.

The entirety of cellular division can be monitored by time lapse image analysis [19]. Monitoring cell lineage analysis reveals a brilliant story of a single cell dividing throughout the intact cell division process. Specifically, the spatiality and temporality of cell cycle, the duration of mitosis, failure of mitosis, and cell death can be analyzed at the single cell level. Above all, MI analysis is an efficient, effective, and resourceful method for understanding molecular dynamics of specific cells and administration of drugs and chemical agents. In this chapter, we will introduce traditional mitotic index analysis with Giemsa-stained metaphase chromosome analysis, fluorescence signal-based microscopic analysis, automated imaging, and flow cytometer analysis against pHH3.

2 Materials

2.1 Cell Culture Solutions, Cell Treatment Buffer

1. Appropriate cell culture media such as MEM, alphaMEM, DMEM, and RPMI supplemented with fetal bovine serum (FBS) and antibiotics. Heat inactivation of FBS at 56 °C for 20 min is recommended.
2. Phosphate-buffered saline (PBS).
3. 10 μg/mL colcemid solution in PBS.
4. Trypsin-EDTA (0.05% or 0.25% trypsin/EDTA-4Na in PBS).
5. Propidium iodide (PI) solution in PBS. Prepare stock solution in 10 μg/ml in PBS.
6. 5% Giemsa solution. Mix 2.5 ml of Giemsa stain with 47.5 ml Gurr buffer solution.
7. 100% ethanol solution.
8. 4% (w/v) paraformaldehyde in PBS solution.
9. 0.2% (v/v) Triton X-100 in PBS solution.
10. 10% goat serum in PBS solution.
11. Rabbit polyclonal anti-phospho-histone H3 (Ser10) antibody.
12. Alexa 546 conjugated goat anti-rabbit IgG antibody.
13. DAPI in SlowFade.

2.2 Equipment

1. Cell culture facility including biosafety cabinet or clean bench, 37 °C CO_2 controlled incubator, and water bath.
2. Centrifuge, which can rotate 15 ml conical tubes. No need of temperature control or brake.
3. Aspirator to remove supernatant and trap waste.
4. Upright microscope.

5. Fluorescent microscope equipped with CCD camera and software.
6. Slide glass, coverslip, mounting medium.
7. Flow cytometer such as Guava PCA (Millipore).
8. Imaging cytometer such as Tali cytometer (Invitrogen).
9. Time lapse recording capable inverted microscope and camera system.

3 Methods

3.1 Mitotic Index Analysis with Simple Metaphase Spreads

1. Cells should be treated with testing agents of interest to researcher.
2. Collect media, wash PBS, and trypsinize, then transfer detached cells into tubes.
3. Centrifuge 200 g for 5 min. Discard supernatant. Tap tubes and break pellets.
4. Add 3 mL of 75 mM KCl for tube and incubate at 37 °C water bath for 20 min.
5. Add 1 mL of freshly prepared fixative solution.
6. Centrifuge 200 g for 5 min. Discard supernatant. Tap tubes and break pellets.
7. Add 1 mL of freshly prepared fixative solution.
8. Centrifuge 200 g for 5 min. Discard supernatant. Tap tubes and break pellets.
9. Add 1 mL of freshly prepared fixative solution.
10. Centrifuge 200 g for 5 min. Discard supernatant. Tap tubes and break pellets.
11. Add 1 mL of freshly prepared fixative solution.
12. Centrifuge 200 g for 5 min. Discard supernatant.
13. Break pellet by pipetting. Drop 30 to 40 μL of fixed cell solution onto clean glass slide for metaphase chromosome spread.
14. Air-dry for 5 min and confirm quality of metaphase chromosome spread.
15. Stain slides with 5% Giemsa solution.
16. Mount coverslip with mounting media.
17. Observe under upright or inverted microscope; typically 20x objective to 40x objective lens should be appropriate to observe mitotic index. Count at least 100 cells and obtain fraction of mitotic cells. Mitotic index = mitotic cell number/total observed cell number (*see* **Note 1**).

3.2 pHH3 Immunocytochemistry Analysis with Fluorescence Microscope

1. Prior to experiment, cells should be plated onto appropriate vessels to observe with microscope.
2. Cells should be treated with testing agents for desired time.
3. Aspirate media and wash with PBS (*see* **Note 2**).
4. Fix cells with 4% paraformaldehyde in PBS for 15 min. Fixation solution should cover all of the cells growing within the coverslip area. For coverslip in 35 mm dish, 1–1.5 ml would be appropriate.
5. Remove paraformaldehyde and wash with PBS.
6. Treat cells with 0.2% Triton X-100 in PBS solution, allowing for permeabilization for 10 min.
7. Remove Triton X-100 and add 10% goat serum for blocking. Apply parafilm to cover cells. For coverslip in dish, 50 μL of 10% goat serum solution should be sufficient.

   Blocking for 30 min at room temperature or 4 °C for overnight.
8. Add pHH3 antibody in 1:500 dilution in 10% goat serum. Apply parafilm to cover cells. 20 μL for square coverslip is an appropriate size for covering.
9. Incubate in 37 °C incubator for 1 h.
10. Wash with PBS three times for 5 min each time.
11. Add fluorescence conjugated secondary antibody in 1:500 dilution in 10% goat serum. Apply parafilm to cover cells. 20 μL for square coverslip is an appropriate size for covering. Incubate in 37 °C incubator for 1 h.
12. Wash with PBS three times for 5 min each.
13. Add mounting media with DAPI and seal with coverslip.
14. Observe and take images of cells with fluorescent microscope equipped with appropriate filters. Cellular nuclei are stained blue with DAPI. pHH3 is stained red. Pan nuclear stain should be observed for positive cells. Two images of DNA and pHH3 need to be merged in appropriate photo software (Fig. 1).
15. Obtain total cell number and pHH3 positive cell population. After counting at least 100 cells, obtain fraction of cell positive with pHH3 to get mitotic index (*see* **Note 3**).

3.3 pHH3 Imaging or Flow Cytometer Analysis for High-Throughput Analysis

1. Prior to experiment, cells should be plated onto appropriate vessels such as dishes or flasks.
2. Cells should be treated with testing agents of interest to researcher.
3. Collect media, PBS, and trypsinized cells into 15 ml conical tube.

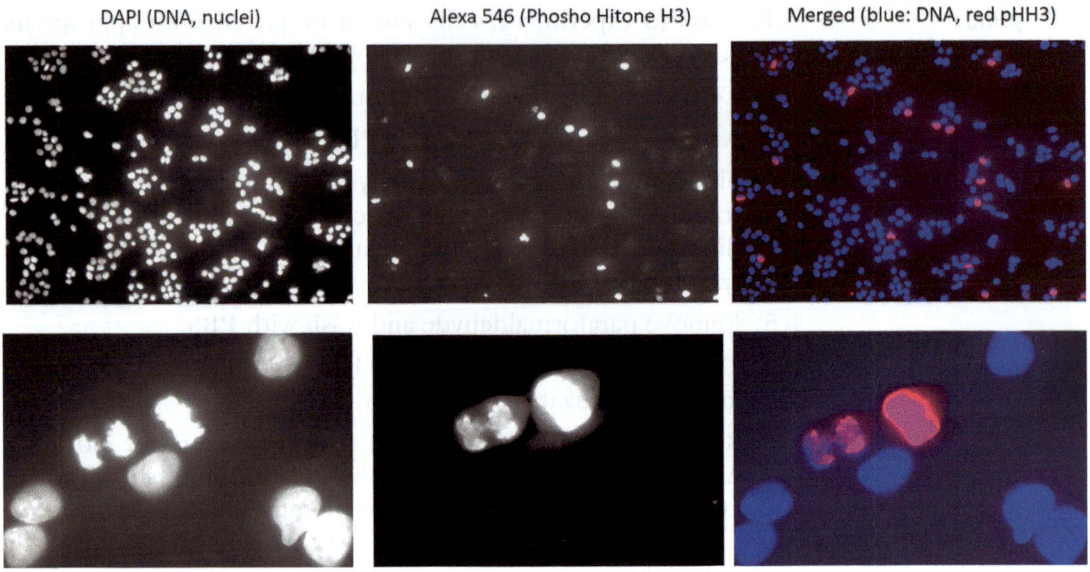

Fig. 1 Indirect fluorescent immunocytochemistry staining of phospho-histone 3 serine 10 (pHH3-S10). Cells in M-phase were stained in red. DNA was counter stained with DAPI. Images were obtained in 20x objective lens

4. Centrifuge samples (300 g, 5 min). Discard supernatant and add 1 ml of PBS and break cell pellet.
5. Centrifuge samples (300 g, 5 min). Discard supernatant and add 0.5 ml of PBS and break cell pellet.
6. Add 1 ml of 100% ethanol in dropwise fashion. Mix well.
7. Cells should be stored in −20 °C for at least overnight length of time. Cells can be stored for weeks and (potentially) months in −20 °C.
8. Centrifuge 300 g for 5 min. Discard supernatant. Tap tubes and break pellets.
9. Add 1 mL of PBS and centrifuge 200 g for 5 min. Discard supernatant. Tap tubes and break pellets.
10. Add 0.5 mL of 0.2% Triton X-100 in PBS and incubate at room temperature for 10 min, and then centrifuge 200 g for 5 min. Discard supernatant. Tap tubes and break pellets.
11. Add 100 μL of 10% goat serum for blocking for 1 h at room temperature.
12. Add 0.25 μL of anti-phospho-HH3-S10 antibody. Incubate for 1 h at 37 °C (*see* **Note 4**).
13. Add 1 mL of PBS and centrifuge 200 g for 5 min. Discard supernatant. Tap tubes and break pellets.
14. Add 1 mL of PBS and centrifuge 200 g for 5 min. Discard supernatant. Tap tubes and break pellets.

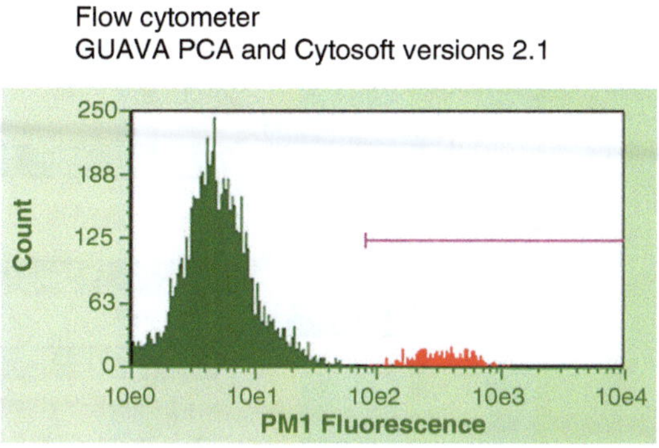

Fig. 2 Flow cytometer analysis of phosphor-histone H3. PM1 (red channel) represents how much red signal per each cell. Green population is phospho-histone H3 negative. Red population is positive. Cytometer software calculates a fraction of red signal as 0.05

15. Add 0.25 μL of Alexa Fluor conjugated anti-rabbit IgG antibody in 100 μl of 10% goat serum. Incubate for 1 h at 37 °C.
16. Add 1 mL of PBS and centrifuge 200 g for 5 min. Discard supernatant. Tap tubes and break pellets.
17. Add 1 mL of PBS and centrifuge 200 g for 5 min. Discard supernatant. Tap tubes and break pellets.
18. Add appropriate amount of PBS to dilute cells in optimum concentrations.
19. Apply cells into cytometer and analyze cells. For flow cytometer, we recommend minimum 10,000 for analysis. Millipore Guava PCA is recommended. For imaging cytometer, we use Invitrogen Tali imaging cytometer. You need to set number of images to analyze rather than number of cells (Fig. 2).

3.4 Time Lapse Image Analysis for Mitosis Event at Single Cell Level

1. Cells should be plated on dish or flask prior experiments. A few hours or overnight pre-plating may be ideal.
2. Add testing agents and place on inverted microscope with temperature regulation and video camera (*see* **Note 5**).
3. Time lapse images should be taken every 1–10 min for desired duration.
4. Images need to be converted to movie by Timelapse conversion software. If images are taken in 5-min interval, frame rate should be set for 12 fps. Therefore, one second of movie will correlate to one hour of real time (*see* **Note 6**).
5. Track each cell and record when cells divide. Record the time of division and illustrate cell lineage (Fig. 3).

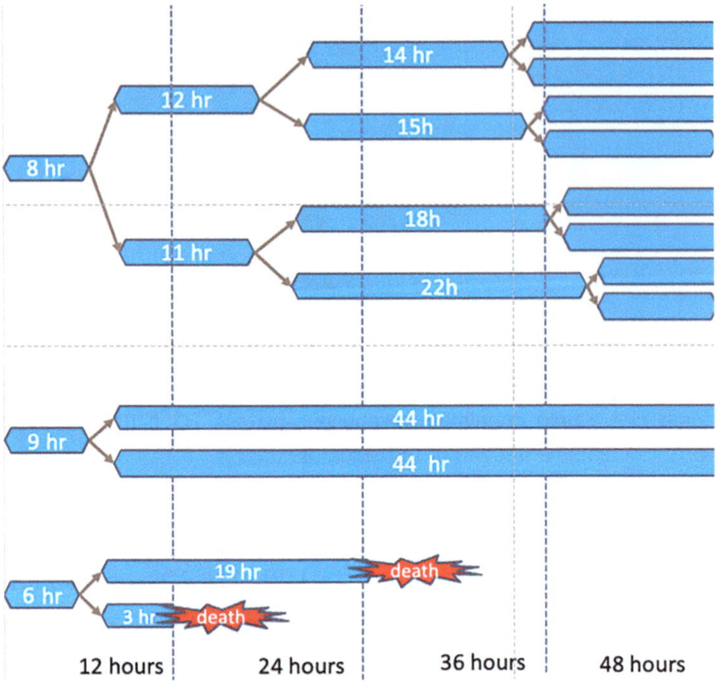

Fig. 3 Cell division lineage trace. Once movie starts, single cells are tracked for their division. We set the time lapse as 1 second; therefore, 1 hour of time lapse movie is readily available to depict the cell dividing. Record the time of division and keep tracking for 1–4 h. Cells may stop dividing due to various factors or die due to apoptosis

4 Note

1. Mitotic index analysis is relatively straightforward while providing substantial data of cellular proliferation activity. It can be combined with other proliferation markers such as BrdU/EdU staining and Ki67 staining to increase understanding of cell cycle progression and checkpoint. Toxic agent exposure decreases mitotic index. However, if cells have deficiency in checkpoint, mitotic index may not be affected as expected.

   Chemicals arresting cells in mitosis such as colcemid, paclitaxel, or nocodazole can be used for a positive control.

2. Mitotic cells are often loosely attached or detached from cell culture vessels. Fixation and staining process must be carried out with care for fluorescence immunocytochemistry.

3. Cell lineage analysis can be combined with other reporter system such as DNA damage protein fused GFP proteins and cell cycle marker proteins fused with fluorescence.

4. Optimize concentration of primary antibody. Signal and noise ratio can be improved with higher concentration of primary antibody.
5. If inverted microscope is placed in CO_2 incubator, or if inverted microscope is equipped with humidifier and CO_2 gas line, no special care is necessary. If inverted microscope is placed in warm room without CO_2 regulation or just heating stage, add 10–25 mM HEPES buffer to maintain pH during time lapse imaging.
6. It may be difficult to store and handle large volumes of images depending on computer CPU power. Normal mitosis event may complete within 1 h. Usually a 5-min interval may be sufficient. AutoKam software can control up to 3 cameras by one computer and take images with specific interval [20].

References

1. Howard A, Pelc S (1953) Synthesis of deoxyribonucleic acid in normal and irradiated cells and its relation to chromosome breakage. Heredity 6(supple):261–273
2. Hoffman JG (1947) Wright's hypothesis: its relation to volume growth of tissue cells and mitotic index. Science 106(2754):343–344. https://doi.org/10.1126/science.106.2754.343-a
3. Hoffman JG (1949) Theory of the mitotic index and its application to tissue growth measurement. Bull Math Biophys 11(2):139–144. https://doi.org/10.1007/BF02477499
4. Kim JY, Jeong HS, Chung T, Kim M, Lee JH, Jung WH, Koo JS (2017) The value of phosphohistone H3 as a proliferation marker for evaluating invasive breast cancers: a comparative study with Ki67. Oncotarget. 8(39): 65064–65076. https://doi.org/10.18632/oncotarget.17775
5. Medri L, Volpi A, Nanni O, Vecci AM, Mangia A, Schittulli F, Padovani F, Giunchi DC, Zito A, Amadori D, Paradiso A, Silvestrini R (2003) Prognostic relevance of mitotic activity in patients with node-negative breast cancer. Mod Pathol 16(11):1067–1075. https://doi.org/10.1097/01.MP.0000093625.20366.9D
6. Wilson L (1975) Action of drugs on microtubules. Life Sci 17(3):303–309. https://doi.org/10.1016/0024-3205(75)90476-2
7. Eigsti OJ (1938) A cytological study of colchicine effects in the induction of polyploidy in plants. Proceedings of the National Academy of Sciences of the United States of America. 24: 56–63. https://doi.org/10.1073/pnas.24.2.56
8. Davidson D, Macleod RD (1966) Changes in mitotic indices in roots of vicia faba I. Antagonistic Effects of Colchicine and Iaa. Chromosoma. 18(3):421. https://doi.org/10.1007/Bf00332546
9. Dewey WC, Miller HH (1969) X-ray induction of chromatid exchanges in mitotic and G1 Chinese hamster cells pretreated with Colcemid. Exp Cell Res 57(1):63–70
10. Alderton GK, Joenje H, Varon R, Borglum AD, Jeggo PA, O'Driscoll M (2004) Seckel syndrome exhibits cellular features demonstrating defects in the ATR-signalling pathway 22. HumMolGenet 13(24):3127–3138
11. Rossi L, Laas E, Mallon P, Vincent-Salomon A, Guinebretiere JM, Lerebours F, Rouzier R, Pierga JY, Reyal F (2015) Prognostic impact of discrepant Ki67 and mitotic index on hormone receptor-positive, HER2-negative breast carcinoma. Br J Cancer 113(7):996–1002. https://doi.org/10.1038/bjc.2015.239
12. Chopra DP, Wilkoff LJ (1977) Induction of hyperplasia and anaplasia by carcinogens in organ cultures of mouse prostate. In Vitro 13(4):260–267. https://doi.org/10.1007/BF02615084
13. Goto H, Tomono Y, Ajiro K, Kosako H, Fujita M, Sakurai M, Okawa K, Iwamatsu A, Okigaki T, Takahashi T, Inagaki M (1999) Identification of a novel phosphorylation site on histone H3 coupled with mitotic chromosome condensation. J Biol Chem 274(36): 25543–25549. https://doi.org/10.1074/jbc.274.36.25543
14. Hendzel MJ, Wei Y, Mancini MA, Van Hooser A, Ranalli T, Brinkley BR, Bazett-Jones DP, Allis CD (1997) Mitosis-specific

phosphorylation of histone H3 initiates primarily within pericentromeric heterochromatin during G2 and spreads in an ordered fashion coincident with mitotic chromosome condensation. Chromosoma 106(6):348–360. https://doi.org/10.1007/s004120050256

15. Paulson JR, Taylor SS (1982) Phosphorylation of histones 1 and 3 and nonhistone high mobility group 14 by an endogenous kinase in HeLa metaphase chromosomes. J Biol Chem 257(11):6064–6072
16. Gurley LR, D'Anna JA, Barham SS, Deaven LL, Tobey RA (1978) Histone phosphorylation and chromatin structure during mitosis in Chinese hamster cells. Eur J Biochem 84(1):1–15. https://doi.org/10.1111/j.1432-1033.1978.tb12135.x
17. Wei Y, Mizzen CA, Cook RG, Gorovsky MA, Allis CD (1998) Phosphorylation of histone H3 at serine 10 is correlated with chromosome condensation during mitosis and meiosis in Tetrahymena. Proc Natl Acad Sci U S A 95(13):7480–7484. https://doi.org/10.1073/pnas.95.13.7480
18. Sawicka A, Seiser C (2012) Histone H3 phosphorylation - a versatile chromatin modification for different occasions. Biochimie 94(11):2193–2201. https://doi.org/10.1016/j.biochi.2012.04.018
19. Forrester HB, Vidair CA, Albright N, Ling CC, Dewey WC (1999) Using computerized video time lapse for quantifying cell death of X-irradiated rat embryo cells transfected with c-myc or c-ha-ras. Cancer Res 59(4):931–939
20. Lynch AE, Triajianto J, Routledge E (2014) Low-cost motility tracking system (LOCOMOTIS) for time-lapse microscopy applications and cell visualisation. PLoS One 9(8):e103547. https://doi.org/10.1371/journal.pone.0103547

Chapter 4

Preparation of Mitotic Cells for Fluorescence Microscopy

Yuji Tanno

Abstract

Cell cycle-dependent regulation of chromosome is a dynamic event. After replication in S phase, sister chromatids show dynamic behavior including condensation, alignment, and segregation in M phase. These beautiful behaviors of chromosomes observed through the microscope have fascinated people since more than 100 years ago, and now we can sketch the dynamics of regulatory proteins and their posttranscriptional modifications through the fluorescent microscope. The purpose of this chapter is describing the basic methods of immunofluorescence analysis of mitotic cells and chromosomes. Besides, the key ideas for improving the preparation of the specimen are also described. Because the characteristic of the proteins of your interest differs one by one, modifying the method might cause the crucial improvement in the observation.

Key words Mitosis, Chromosome, Immunofluorescence, Immunostaining, Microscopy, Chromatin, Antibody

1 Introduction

Human cells transmit their genetic information to progeny cells by replicating DNA in S phase and partitioning the duplicated DNA into daughter cells in M phase. The behavior of cells and chromosomes in M phase are extraordinary coordinated, dynamic, and beautiful processes. These processes include nuclear envelope breakdown, chromosome condensation which is followed by alignment at the cell equator, and segregation into daughter cells, and cytokinesis [1]. Immunofluorescence microscopy is a technique which has been used for examining the regulatory mechanism of chromosome behavior in mitosis such as chromosome condensation, chromatin structure and posttranscriptional modification of regulatory proteins, kinetochore formation at centromeric chromatin, organization of spindle microtubules and their attachment to kinetochore.

Eisuke Gotoh (ed.), *Chromosome Analysis: Methods and Protocols*, Methods in Molecular Biology, vol. 2519,
https://doi.org/10.1007/978-1-0716-2433-3_4,

In this chapter, methods of sample preparation for immunofluorescent analysis of mitotic chromosome are described. Depending on the characteristic of target protein or posttranslational modification, you can examine and choose the appropriate method. For example, the whole structure of the cells should be appropriately retained when you would like to test the alignment of chromosome, spindle microtubule structure [2]. On the other hand, to observe individual chromosomes for careful examination of the structural proteins on them, cells should be swollen by hyposmolality and spun onto the glass slide by cytospin to perform immunostaining of chromosome spread [3]. In the case that the amount of target molecule is quite small or the localization of the molecule is limited, the signal could be hidden by the background noise. The cross-reaction of the antibody against other molecules also could be the misdetection. In such cases, changing the method might critically improve the detection of the target. For example, signal of phosphorylated proteins at kinetochore is frequently covered by other signals, and the signal is unclear with basic method of cell fixation. On the other hand, extraction of the cytoplasmic signal or preparation of the chromosome spread before fixing cells might promote the proper access of antibody specifically to the target protein and cause better signal [4, 5].

We describe the basic procedures of the immunostaining of mitotic cells. First of all, the immunostaining method for cells fixed on coverslip is introduced. In this method, the whole structure and the content of the cells is moderately maintained. Next, the methods with the extraction of the cytoplasm and proteins are weakly associated with chromatin. In addition, the method for preparing the chromosome spread with hypotonic treatment and cytocentrifuge is explained. Furthermore, the tips for the sample preparation of the cells and cell cycle synchronization are shown. The notes for considering the better procedure and troubleshooting and the explanation of key parameters for the improvement of the sample preparation are added for each technique.

2 Materials

2.1 Solutions

1. 1x phosphate-buffered saline (PBS (−)): 137 mM NaCl, 2.7 mM KCl, 10 mM Na_2HPO_4, 1.8 mM KH_2PO_4, dissolved in deionized water.
2. Hypotonic buffer: 25% PBS (−), diluted with deionized water.
3. 10% Triton X-100: Dilute 100% Triton X-100 with to 10%, 1x PBS (−).
4. Permeabilization buffer (0.2% Triton X-100): Dilute 10% Triton X-100 to 0.2% with PBS (−).

5. Fixative.
   1. 4% PFA, 1.4 mM NaOH: Dissolve in PBS (−). Incubate the mixture for 30 min at 65 °C.
   2. 2% PFA, 1% PFA: Dilute 4% PFA in PBS (−) to given concentration, respectively.
   3. Extractive fixative: 0.2% Triton, 2% PFA, diluted in PBS (−).
   4. Blocking buffer/antibody dilution buffer: 3% Bovine serum albumin (BSA) dissolved in PBS (−).
   5. Mounting buffer: 70% glycerol in1x PBS (−).

2.2 Tools for Immunofluorescence Staining

1. Cell culture tools: 12-Well cell culture plate, φ3.5 cm cell culture dish, φ12 mm coverslip, cell culture medium supplemented with serum and antibiotics (e.g., Dulbecco's Modified Eagle Medium with 10% fetal bovine serum and 100 units/mL penicillin and100 μg/mL streptomycin), incubator.
2. Cell adhesive: Fibronectin, Cell-Tak (Corning).
3. Antibodies and DNA stain: A pair of primary antibody against your target protein (e.g., anti-CENP-C antibody (MBL, PD030) used at 1:1000 dilution) and secondary antibody (e.g., Alexa Fluor 647 anti-guinea pig antibody (Thermo Fisher, A21450) for the above-indicated CENP-C antibody and at 1:1000 dilution), HOECHST 33342 for counterstaining DNA/nucleus.
4. Other equipment: Forceps, parafilm, nail topcoat, liquid blocker pen, water bath.
5. Cytocentrifuge: Cytospin 4 (Thermo Fisher Scientific), cytoglass, cytofunnel.

2.3 Reagent for Synchronization of Cell Cycle

1. 100 mM thymidine: Thymidine dissolved in PBS (−), filtrate with 0.22 μm filter.
2. 1 mM nocodazole: Dissolved in dimethyl sulfoxide (DMSO).
3. 5 μM Taxol: Dissolved in DMSO.
4. 20 μM MG132: Dissolved in DMSO.

3 Methods

3.1 Immunostaining of Cells Fixed on Coverslip

This technique is the basic method for observing the structure and the location of the chromosomes and organelles in a cell; thus, you can discuss about the presence or absence and the localization of the chromatin and its regulatory factors under a microscope [6–8]:

1. Place a coverslip in a well of 12-well cell culture plate (*see* **Note 1**).
2. Seed cells at 50% confluency in the well (*see* **Note 2**).

3. Proliferate the cells in an incubator for 16 h (*see* **Note 3**).
4. Aspirate the medium and wash the cells with 1 mL of PBS (−) (*see* **Note 4**).
5. Aspirate the PBS (−) and add the 1 mL of 4% PFA in the well (*see* **Note 5**).
6. Incubate for 7 min, and at this stage cells are fixed (*see* **Note 6**).
7. Aspirate the PFA and wash once with 1 mL of PBS (−).
8. Add 1 mL of 0.2% Triton X-100 in the well (*see* **Note 7**).
9. Incubate for 10 min at room temperature.
10. Aspirate the 0.2% Triton X-100 and wash once with 1 mL of PBS (−).
11. Add 1 mL of blocking buffer and incubate for 15 min.
12. Quickly transfer the coverslip on the parafilm. The cell-attached side should be upward (*see* **Note 8**).
13. Put 100 μL of the primary antibody solution on the coverslip (*see* **Note 9**).
14. Put the cover of the cell culture plate on it to prevent the drying of the solution, and incubate for 1 h.
15. In a new 12-well plate, put 1 mL of PBS (−) in a well and transfer the coverslip into the well. The surface with cell should be upward.
16. Aspirate the PBS (−) and wash the coverslip twice with PBS (−). At this washing step, incubate the coverslip with PBS (−) for 5 min.
17. Quickly transfer the coverslip on the parafilm (*see* **Note 8**).
18. Put 100 μL of the secondary antibody solution on the coverslip (*see* **Notes 9 and 10**).
19. Put the cover of the cell culture plate on it to prevent the drying of the solution, and cover with cooking foil to shield them from the light, and then incubate for 1 h.
20. In a 12-well plate, put 1 mL of PBS (−) in a well and transfer the coverslip into the well. The side with cell should be upward (*see* **Note 11**).
21. Aspirate the PBS (−) and wash the coverslip three times with PBS (−). At this washing step, incubate the coverslip with PBS (−) for 10 min.
22. On a glass slide, put the 100 μL of mounting buffer and put the coverslip (the cell-side should face to the glass slide) (*see* **Note 12**).
23. Gently push the coverslip with forceps to remove excess amount of the mounting buffer between coverslip and glass slide and wipe off the extra leaked mounting buffer (*see* **Note 13**).

24. Seal the edge of the coverslip and surrounding region of the glass slide with nail topcoat (*see* **Note 14**).
25. The sample is ready for microscopic analysis (*see* **Note 15**).

3.2 Cell Extraction Before the Fixation

The basic coverslip method described in Subheading 3.1 is a simple and easy one; however, it is not suitable for some target protein or posttranscriptional modification signals. Some proteins bind to the chromatin but also exist abundantly in cytoplasm, and cytoplasmic signal of the immunofluorescence might mask and diminish the signal of the target localized at particular region (e.g., chromatin, centromere, and telomere). Also, numerous proteins are phosphorylated in mitosis, so the researchers generate phospho-specific antibodies against a certain protein of interest, however it could bind nonspecifically to non-phosphorylated form of the protein of interest and also to other phosphorylated proteins. The extraction of the cells with detergent and/or physical force by centrifugation causes the exposure of the chromatin structure to a greater extent, and the signal of interest could be more efficiently detected. Here, three methods of cell extraction are described: (1) the treatment of the cells with Triton X-100 before the fixation [8, 9], (2) the technique which is called cytospin method—cells are spun onto the glass slide by centrifugation and soft structure of the cells are extracted [6, 8]—and (3) the hypotonic treatment of the cells with the PBS (−) diluted with water (this technique is similar to the one used for the preparation of the chromosome sample for Giemsa staining). Cells treated with hypotonic buffer are then processed for the cytocentrifuge; thus, this method causes the most vigorous extraction for the cell [3, 8].

3.2.1 Extraction of the Cells During the Fixation

1. Place a coverslip in a well of 12-well cell culture plate (*see* **Note 1**).
2. Seed cells at 50% confluency in the well (*see* **Note 2**).
3. Proliferate the cells in an incubator for 16 h (*see* **Note 3**).
4. Aspirate the medium and wash the cells with 1 mL of PBS (−) (*see* **Note 4**).
5. Add 1 mL of 1% PFA/0.2% Triton X-100 in the well (*see* **Note 16**).
6. The following step is same as 3.1 from the step 6.
   Incubate for 7 min at room temperature (*see* **Note 6**).
7. Aspirate the medium and wash the cells with 1 ml PBS (−).
8. Add 1 mL of 0.2% Triton X-100 in the well (*see* **Note** 7).
9. Incubate for 10 min at room temperature.
10. Aspirate the 0.2% Triton X-100 and wash once with 1 mL PBS (−).

11. Add 1 mL of blocking buffer and incubate for 15 min.
12. Quickly transfer the coverslip on the parafilm. The surface with cell-attached side should be upward (*see* **Note 8**).
13. Put 100 μL of the primary antibody solution on the coverslip (*see* **Note 9**).
14. Put the cover of the cell culture plate on it to prevent the drying of the solution, and incubate for 1 h.
15. In a new 12-well plate, put 1 ml of PBS (−) in a well and transfer the coverslip into the well. The surface with cell should be upward.
16. Aspirate the PBS (−) and wash the coverslip twice with PBS (−). At this washing step, incubate the coverslip with PBS (−) for 5 min.
17. Quickly transfer the coverslip on the parafilm (*see* **Note 8**).
18. Put 100 μL of the secondary antibody solution on the coverslip (*see* **Notes 9 and 10**).
19. Put the cover of the cell culture plate on it to prevent the drying of the solution, and also cover with cooking foil to shield them from the light, and then incubate for 1 h.
20. In a 12-well plate, put 1 mL of PBS (−) in a well and transfer the coverslip into the well. The surface with cell should be upward (*see* **Note 11**).
21. Aspirate the PBS (−) and wash the coverslip three times with PBS (−). At this washing step, incubate the coverslip with PBS (−) for 10 min.
22. On a glass slide, put the 100 μL of mounting buffer and put the coverslip (the cell-side should face to the glass slide) (*see* **Note 12**).
23. Gently push the coverslip with forceps to remove excess amount of the mounting buffer between coverslip and glass slide and wipe off the leaked mounting buffer (*see* **Note 13**).
24. Coat the edge of the coverslip and surrounding region of the glass slide with nail topcoat (*see* **Note 14**).
25. The sample is ready for microscopic analysis (*see* **Note 15**).

3.2.2 Extraction of the Cells Before the Fixation

1. Place a coverslip in a well of 12-well cell culture plate (*see* **Note 1**).
2. Seed cells at 50% confluency in the well (*see* **Note 2**).
3. Proliferate the cells in an incubator for 16 h (*see* **Note 3**).
4. Aspirate the medium and wash the cells with 1 mL of PBS (−) (*see* **Note 4**).
5. Add 1 ml of 0.2% Triton X-100 and incubate for 2 min at room temperature.

6. Gently aspirate the 0.2% Triton X-100 and add 2% PFA (*see* **Note 17**).
7. The following step is same as 3.1 from the step 6.
   Incubate for 7 min at room temperature (*see* **Note 6**).
8. Aspirate the PFA and wash once with 1 mL of PBS (−).
9. Add 1 mL of 0.2% Triton X-100 in the well (*see* **Note 7**).
10. Incubate for 10 min at room temperature.
11. Aspirate the 0.2% Triton X-100 and wash once with 1 mL of PBS (−).
12. Add 1 mL of blocking buffer and incubate for 15 min.
13. Quickly transfer the coverslip on the parafilm. The surface with cell-attached side should be upward (*see* **Note 8**).
14. Put 100 μL of the primary antibody solution on the coverslip (*see* **Note 9**).
15. Put the cover of the cell culture plate on it to prevent the drying of the solution, and incubate for 1 h.
16. In a new 12-well plate, put 1 mL of PBS (−) in a well and transfer the coverslip into the well. The surface with cell should be upward.
17. Aspirate the PBS (−) and wash the coverslip twice with PBS (−). At this washing step, incubate the coverslip with PBS (−) for 5 min.
18. Quickly transfer the coverslip on the parafilm (*see* **Note 8**).
19. Put 100 μL of the secondary antibody solution on the coverslip (*see* **Notes 9 and 10**).
20. Put the cover of the cell culture plate on it to prevent the drying of the solution, and cover with cooking foil to shield them from the light, and then incubate for 1 h.
21. In a 12-well plate, put 1 mL of PBS (−) in a well and transfer the coverslip into the well. The surface with cell should be upward (*see* **Note 11**).
22. Aspirate the PBS (−) and wash the coverslip three times with PBS (−). At this washing step, incubate the coverslip with PBS (−) for 10 min.
23. On a glass slide, put the 100 μL of mounting buffer and put the coverslip (the cell-side should face to the glass slide) (*see* **Note 12**).
24. Gently push the coverslip with forceps to remove excess amount of the mounting buffer between coverslip and glass slide and wipe off the leaked mounting buffer (*see* **Note 13**).
25. Coat the edge of the coverslip and surrounding region of the glass slide with nail topcoat (*see* **Note 14**).
26. The sample is ready for microscopic analysis (*see* **Note 15**).

3.2.3 Hypotonic Treatment of Cells and Cytocentrifugation

1. Seed the cells at 50% confluency in a 6-well cell culture plate.
2. Proliferate the cells for 16 h in an incubator.
3. Harvest the cells from the culture dish with trypsin.
4. Wash the cells with 1 mL of PBS (−) with centrifugation at 200 g for 5 min.
5. Aspirate the PBS (−) and resuspend the cell pellet with the hypotonic buffer.
6. Incubate the cell suspension for 15 min at 37 °C using water bath (*see* **Note 18**).
7. During incubation, mark the circle with liquid blocker at the region surrounding the sample position on a cytoslide, and hold it and a cytofunnel with a sample chamber of the cytospin (*see* **Note 19**).
8. Resuspend the cells by pipetting and apply 100 μL into a cytofunnel.
9. Perform cytocentrifugation at 900 rpm for 5 min (*see* **Note 20**).
10. During cytocentrifugation, add 2% PFA in a container for soaking the cytoslide.
11. Disassemble the cytofunnel and transfer the cytoslide into the container with PFA as quickly as possible to prevent the drying of the sample.
12. Incubate for 7 min at room temperature.
13. By decantation, remove the PFA from the container and wash once with PBS (−).
14. Add the blocking buffer and incubate for 15 min.
15. Place the cytoglass on the bench and remove the buffer on the cells and add 100 μL of the primary antibody solution.
16. Put the cover to prevent the drying of the solution, and incubate for 1 h.
17. Wash the cytoglass with PBS (−) twice in a container. At this washing step, incubate the cytoglass with PBS (−) for 5 min.
18. Place the cytoglass on the bench and remove the buffer on the cells and add 100 μL of the secondary antibody solution (*see* **Note 10**).
19. Put the cover of the cell culture plate on it to prevent the drying of the solution, and also cover with cooking foil to shield them from the light, and then incubate for 1 h.
20. Wash the cytoglass with PBS (−) three times in a container. At this washing step, incubate the cytoglass with PBS (−) for 10 min.

21. On a sample position surrounded by the liquid blocker, put the 100 μL of mounting buffer and put the coverslip (*see* **Note 12**).
22. Gently push the coverslip with forceps to remove excess amount of the mounting buffer between coverslip and cytoglass and wipe off the leaked mounting buffer (*see* **Note 13**).
23. Coat the edge of the coverslip and surrounding region of the cytoglass with nail topcoat. This coating prevents the drying of the sample (*see* **Note 14**).
24. The sample is ready for microscopic analysis (*see* **Note 15**).

3.3 Coating of Coverslip with Reagents to Assist the Attachment of Cells

Here, the techniques to enhance the attachment of cells on coverslip are described. Fibronectin supports the attachment of weak adhesive cells, and Cell-Tak enables the attachment of even floating cells such as cells dissociated from the culture dish or blood cells [9].

3.3.1 Coating Coverslip with Fibronectin

1. Put a coverslip in a well of 12-well cell culture plate.
2. Soak a coverslip in 300 μL of the 50 μg/mL fibronectin solution, and incubate for 1 h at room temperature.
3. Wash twice with 1 mL PBS (−).
4. Seed the cells with cell culture medium.

3.3.2 Coating of Coverslip with Cell-Tak

1. Put a coverslip in a well of 12-well cell culture plate.
2. Soak in 500 μL of PBS (−) (*see* **Note 21**).
3. Add 14 μg Cell-Tak in a well and mix with PBS (−) by gently shaking the plate.
4. Put the cover of the plate and incubate for 10 min at room temperature.
5. Aspirate the solution and wash three times with PBS (−).
6. Seed the cells with the medium without serum because serum inhibits the attachment of cells with Cell-Tak coat.
7. After incubation for 30 min in a cell culture incubator, change the medium into the cell culture medium containing the serum (e.g., 10% fetal bovine serum).

3.4 Synchronization of Cell Cycle and Accumulation of Mitotic Cells

In this section, methods to synchronize cell cycle into M phase are described. Doubling time of human cell line is generally about 24 h, and M phase is started with nuclear envelope breakdown and concurrent chromosome condensation and finished with cytokinesis within about 1 h. Thus, the population of the cells in M phase is below 5% and it is difficult to analyze enough number of cells in specific stage of M phase. To overcome this, there are several techniques to synchronize the cells into M phase and there are chemicals used to synchronize at specific stage in M phase.

3.4.1 Thymidine Block and Release

Thymidine blocks the synthesis of DNA replication and block the cell cycle at mid-S phase. The cells restart the DNA synthesis and complete the replication after deprive of thymidine. Thus, certain time after the release from thymidine block, cells progress into G2, M, and G1 phase. This so-called thymidine block and release method is used for the synchronization of cells at various cell cycle stages. In the case of HeLa cells, cells transit to M phase about 10 h after release from thymidine block. Here, the double thymidine block and release method, which enables highly effective synchronization of the cells, is described. Single thymidine block and release also can synchronize the cell cycle at some extent so it is brief and useful as well.

1. Seed the cells at 25% confluency in the culture dish.
2. Add 1:50 volume of 100 mM thymidine into the medium; thus, the thymidine concentration would be 2 mM (*see* **Note 22**).
3. Incubate the cells for 16 h in an incubator.
4. Aspirate the medium and wash the cells with PBS (−) of a half volume of the culture medium.
5. Add the medium and incubate for 12 h in the incubator.
6. Add 2% volume of 100 mM thymidine into the medium.
7. Incubate the cells for 12 h in the incubator.
8. Aspirate the medium and wash the cells with PBS (−) of a half volume of the culture medium.
9. Add the medium and incubate in the incubator.
10. In the case of HeLa cells, mitotic cells are accumulated from 10 h after second thymidine release (*see* **Note 23**).

3.4.2 Prometaphase Arrest with Nocodazole or Taxol

Nocodazole depolymerizes the tubulin and prevents the formation of spindle microtubules; hence, cells cannot align the chromosomes [8, 10]. Taxol stabilizes the tubulin polymer and chromosome movement is constrained [2]. Thus, these drugs arrest the cells at prometaphase of M phase.

1. Seed the cells at 50% confluency.
2. Incubate the cells for 16 h in an incubator.
3. Incubate the cells with 1 μM nocodazole or 5 μM Taxol (*see* **Notes 24 and 25**.)
4. At 1 h after addition of the drugs, you can clearly observe the accumulation of the prometaphase cells. At the appropriate timing, you can process the cells for immunostaining shown in Subheadig 3.1, 3.2, and 3.3.

3.4.3 Metaphase Arrest with MG132

MG132 inhibits proteasome of which activity is required for metaphase-anaphase transition in M phase; thus, this drug causes metaphase arrest [9, 11]. Note that proteasome also acts at G2-M transition as well; hence, also arrest the cell at G2 if cells are treated with MG132 before entrance into M phase.

1. Seed the cells at 50% confluency.
2. Incubate the cells for 16 h in an incubator.
3. Add 1:1000 volume of 20 mM MG132 (*see* **Note 25**).
4. At 1 h after addition of MG132, you can clearly observe the mitotic cells arrested in metaphase. At the appropriate timing, you can immunostain the cells as shown in Subheading 3.1, 3.2, and 3.3.

4 Notes

1. It could be modified by the combination of cell culture dish and the size of coverslips. For example, you can place 3 coverslips in a φ3.5 cm dish. To handle the coverslip, it is convenient to use the thin and flat type of forceps.
2. The cell confluency depends on the length of doubling time. Cells with 90% confluency on the coverslip are suitable for the observation of mitotic cells because majority of cell lines show mitotic rounding and easily detached from the coverslips during the buffer exchange. Thus, surrounding cells can support each other in properly confluent situation, and it is important to seek such condition.
3. As noted in **Note 2**, you can modify the incubation time for finding appropriate condition of the cells.
4. At washing step, just add and aspirate PBS (−) and take care about drying during washing. Although it depends on the cell types, many cell lines show mitotic rounding and loose attachment to the coverslip compared to the stage of interphase. Thus, the washing of cells should be gentle and careful.
5. Concentration of the PFA is a critical factor for immunostaining. In my experience, 4% PFA is suitable for staining spindle microtubules. On the other hand, lower concentration of PFA such as 1% and 2% can lead better access of antibody to the target protein on chromatin, kinetochore. The background signal might be significantly reduced if you use lower concentration of PFA for these proteins. However, microtubules are depolymerized during fixation with lower concentration of PFA because of lower ability of fixative. Thus, finding proper concentration of fixative for your target is crucial.
6. Incubation time could be modified.

7. At this process, cells are permeabilized, and it helps the access of antibody to the target.
8. Sheet the parafilm on the bench or proper size of glass plate. Keep horizontal during the experiment.
9. You can modify the volume of the solution for the surface of coverslip. Sufficient amount should be chosen.
10. HOECHST 33258 could be added if you would like to stain the DNA/chromosome.
11. At this step, the same plate used for the wash of primary antibody could be reused.
12. There is no problem to use the mixture of glycerol in PBS (−) for the mounting buffer. However, you can choose the commercial mounting buffer containing antifade reagent if you have a problem of quenching of the fluorescence under microscopy.
13. For removing the excess buffer, it is convenient to use soft tissue paper.
14. Regarding nail topcoat, smooth type is easier to seal the glasses.
15. Coating with nail topcoat prevents the drying of the sample during storage, and you can keep the sample at 4 °C until if you start observation later. The storable period depends on the target protein and the strength of the fixation.
16. Mixing Triton X-100 in PFA solution have mild effect compared to the pre-extraction before the fixation shown in 3.2.2, but surely effective for removing the cytoplasmic fraction moderately. You can modify the percentage of PFA as in **Note 5** and you can also examine the proper percentage of the Triton X-100 as well.
17. During the treatment with Triton X-100, some types of cell tend to be detached from the coverslips. In such case, Cell-Tak is more effective to attach the cells to coverslip.
18. Some cell types show tolerance to the low osmotic pressure and difficult to be swollen, in such case you can try the lower concentration of PBS (−) in the hypotonic buffer and longer incubation time.
19. The size of the mark with liquid blocker should be larger than the size of the coverslip to be mounted.
20. When the spreading of cells is not enough, you can choose higher speed of centrifugation.
21. PBS (−) is not described in the manual of the Cell-Tak, but quite efficient for coating the glass coverslip because coverslip can be sought in the PBS (−) but not the buffer described in the manual.

22. Some cell types are resistance to thymidine block.
23. You can combine the method with chemical reagent in 3.5.2–3.5.4 for efficient accumulation of the mitotic cells. Especially, the addition of nocodazole at the timing of second thymidine release and 16 h incubation can lead to >90% prometaphase cells.
24. Depolymerization of the spindle microtubules should be examined by the immunofluorescence because insufficient concentration of nocodazole does not disrupt the spindle and mitotic arrest is not occurred.
25. I recommend adding the drug into the PBS (−) and PFA before and during fixative to prevent the washout of the drug and re-enter the cell cycle during the preparation of the sample for immunofluorescence.

Acknowledgments

I appreciate Dr. E. Gotoh for the opportunity of writing this chapter and his fundamental training, which let me recognize about the importance of parameters for sample preparation of biological specimens. I also appreciate Dr. T. Kitajima for his initial education of chromosome study and the immunostaining methods. I would like to show my gratitude to Dr. A. Okabe, Dr. K. Ishiguro, Dr. T. Tsukahara, Dr. T. Honda, Dr. M. Shintomi, Dr. T. Akera, Dr. S. Yamamoto, Dr. Y. Asai, Mr. H. Nakamura, and Mr. H. Susumu for their instruction and discussion about improving the sample preparation for immunofluorescence. I also thank Ms. A. Shiraki, Dr. K. Takakura, Dr. Y. Hosoi, Dr. C. Toyoshima, Dr. M. Hoshi and Dr. Y. Terada for their science education, and Ms. M. Kasuya, Mr. K. Takahashi, Mr. H. Hirakawa and Mr. S. Iida for their kind support.

References

1. Walczak CE, Cai S, Khodjakov A (2010) Mechanisms of chromosome behaviour during mitosis. Nat Rev Mol Cell Biol 11(2):91–102. https://doi.org/10.1038/nrm2832
2. Magidson V, He J, Ault JG, O'Connell CB, Yang N, Tikhonenko I, McEwen BF, Sui H, Khodjakov A (2016) Unattached kinetochores rather than intrakinetochore tension arrest mitosis in taxol-treated cells. J Cell Biol 212(3):307–319. https://doi.org/10.1083/jcb.201412139
3. Hirota T, Lipp JJ, Toh BH, Peters JM (2005) Histone H3 serine 10 phosphorylation by Aurora B causes HP1 dissociation from heterochromatin. Nature 438(7071):1176–1180. https://doi.org/10.1038/nature04254
4. Kitajima TS, Sakuno T, Ishiguro K, Iemura S, Natsume T, Kawashima SA, Watanabe Y (2006) Shugoshin collaborates with protein phosphatase 2A to protect cohesin. Nature 441(7089):46–52. https://doi.org/10.1038/nature04663
5. Dai J, Sullivan BA, Higgins JM (2006) Regulation of mitotic chromosome cohesion by Haspin and Aurora B. Dev Cell 11(5):741–750. https://doi.org/10.1016/j.devcel.2006.09.018

6. Yamagishi Y, Honda T, Tanno Y, Watanabe Y (2010) Two histone marks establish the inner centromere and chromosome bi-orientation. Science 330(6001):239–243. https://doi.org/10.1126/science.1194498
7. Kitajima TS, Hauf S, Ohsugi M, Yamamoto T, Watanabe Y (2005) Human Bub1 defines the persistent cohesion site along the mitotic chromosome by affecting Shugoshin localization. Curr Biol 15(4):353–359. https://doi.org/10.1016/j.cub.2004.12.044
8. Tanno Y, Kitajima TS, Honda T, Ando Y, Ishiguro K, Watanabe Y (2010) Phosphorylation of mammalian Sgo2 by Aurora B recruits PP2A and MCAK to centromeres. Genes Dev 24(19):2169–2179. https://doi.org/10.1101/gad.1945310
9. Asai Y, Fukuchi K, Tanno Y, Koitabashi-Kiyozuka S, Kiyozuka T, Noda Y, Matsumura R, Koizumi T, Watanabe A, Nagata K, Watanabe Y, Terada Y (2019) Aurora B kinase activity is regulated by SET/TAF1 on Sgo2 at the inner centromere. J Cell Biol 218(10):3223–3236. https://doi.org/10.1083/jcb.201811060
10. Zieve GW, Turnbull D, Mullins JM, McIntosh JR (1980) Production of large numbers of mitotic mammalian cells by use of the reversible microtubule inhibitor nocodazole. Nocodazole accumulated mitotic cells. Exp Cell Res 126(2):397–405. https://doi.org/10.1016/0014-4827(80)90279-7
11. Josefsberg LB, Galiani D, Dantes A, Amsterdam A, Dekel N (2000) The proteasome is involved in the first metaphase-to-anaphase transition of meiosis in rat oocytes. Biol Reprod 62(5):1270–1277. https://doi.org/10.1095/biolreprod62.5.1270

Chapter 5

Chemical-Induced Premature Chromosome Condensation Protocol

Eisuke Gotoh

Abstract

Chromosome analysis is one of most fundamental techniques for cytogenetic studies. Chromosomes are conventionally prepared from mitotic cells arrested by colcemid block protocol. Premature chromosome condensation (PCC) technique is an alternative to obtain chromosomes. It was more than half century ago that the first observation of PCC phenomena reported. Since then, cell-fusion-mediated PCC method has been developed and introduced in many fields of chromosome analysis. More than quarter century ago, novel PCC technique using chemical drug has been developed. Afterwards, this simple and efficient drug-induced PCC technique becomes a standard protocol for preparing chromosomes. Thus, it seems to be the good time to introduce PCC technique protocol for the artisans in the field of cytogenetic studies.

Key words Cytogenetic study, Premature chromosome condensation (PCC), Prematurely condensed chromosomes (PCCs), Protein phosphatase inhibitors, Calyculin A (CA), Okadaic acid (OA), Mitotic index (MI), PCC index

1 Introduction

Analysis of chromosomes is widely performed in various fields of cytogenetics, such as heredity genetics, mutagenic studies, oncology, radiation biology, or environmental science. Chromosome analysis has been achieved traditionally by obtaining mitotic phase-arrested chromosomes by colcemid blocking. The established colcemid block protocol is very popular and widely used; however, the artisans have frequently experienced that it is often difficult or even impossible to obtain sufficient numbers of mitotic chromosomes from the cells under certain circumstances—cells growing very slow or even arrested, some kinds of tumor cells or highly damaged cells after exposure of high doses of cell-damaging agents such as ionizing irradiation [1], because these cells cease cell cycle and do not enter in mitosis. An alternative useful method is premature chromosome condensation (PCC, *see* **Note 1**) method

Eisuke Gotoh (ed.), *Chromosome Analysis: Methods and Protocols*, Methods in Molecular Biology, vol. 2519, https://doi.org/10.1007/978-1-0716-2433-3_5,

[2–4] that facilitates the visualization of interphase chromatin as condensed form of chromosomes [5–11]. Therefore, a lot of literatures using this method have been published in cytogenetic studies, particularly in radiation biology and DNA replication/repair analysis [9, 12–25]. PCC has been usually achieved by fusing the recipient interphase cells with mitotic donor cells utilized with either fusogenic virus (e.g., Sendai virus) or polyethylene glycol (PEG): cell fusion-mediated PCC or fusion-PCC [2, 3, 26, 27]. However, fusion PCC has several drawbacks; (1) the fusion efficiency is critically dependent on virus activity, and each lot quality of polyethylene glycol (PEG) or cell viability, in addition resulting PCC index, is often very low, (2) cell fusion procedure is time consuming and technically demanding, and (3) the obtained chromosomes are mixture of mitotic inducer cells and recipient interphase cells; hence, differential replication staining of mitotic cells is often required [26, 28]. As a result, fusion PCC has been utilized in the limited number of laboratories, although it has been recognized as a quite valuable tool for chromosome analysis [5–7, 13–15, 26, 29–31]. PCC induction by chemical agents (drug-induced PCC) was therefore desired, and numbers of approaches have been carried out [32–34]. The first successive drug-induced PCC was introduced about quarter century ago; calyculin A or okadaic acid, specific inhibitors of protein serine/threonine phosphatase, directly induces premature chromosome condensation in mammalian cells in any phase of cell cycle [4, 9, 10]. Drug-induced PCC has several advantages over colcemid block or fusion-PCC: (1) It is very easy to practice by simply substituting colcemid to calyculin A or okadaic acid. (2) Approximately only 30 min incubation with calyculin A induces a sufficient number of chromosomes, that is, much shorter than colcemid block method (it takes usually 2–4 h). Incubation time can be even much shortened to 5–10 min depending on the experimental design [35, 36]. (3) PCC index is usually much higher (>20%) than mitotic index (usually 1–2%, but would be often much lower) that makes the observation and chromosome analysis much easier [4]. Therefore, drug-induced PCC have been becoming popular, and many articles using the technique have been published [9–11, 18, 19, 21, 24, 35, 37–46], such as biological radiation dosimeter [25, 46, 47]. In addition, the drug-induced PCC has an outstanding merit that allows the interphase chromatin (i.e., G1- S- and G2-phase chromatin) to be visualized as condensed form of chromosome structure [4, 28, 48]. Hence, the drug-induced PCC method is a useful way not only as an alternative method for mitotic chromosome analysis but also as new analytical tools for interphase chromatin cytogenetics [11, 20, 28]. In this chapter, the detailed protocol of drug-induced PCC will be described. The protocol is very simple as easily realized by the unskilled person in chromosome preparation, and it is hopefully that the drug-induced PCC will be used in wider ranges in cytogenetics and related fields (*see* **Note 2**).

2 Materials

1. Calyculin A (*see* **Note 3**): Hundred μg of calyculin A, lyophilized in a vial supplied from the supplier, dissolving with 1 mL of solvent (100% DMSO or 100% ethanol, calyculin A does not dissolve in water) will make ~100 μM stock solution (MW of calyculin A is 1009.17). Aliquot 100 μL in a microcentrifuge tube and stored at −20 °C and stable at least 3 months (*see* **Notes 3 and 4**).
2. 0.075 M KCl: Dissolve 7.45 g KCl (M.W. 74.55) in 100 mL Milli-Q water or RO (reverse osmotic) water to make a 1 M KCl stock solution. Take 7.5 ml of the 1 M stock solution and make it up to 100 mL with Milli-Q water or RO water to obtain a 0.075 M KCl working solution. Usually, it does not need sterilizing.
3. Carnoy's fixative (methanol: glacial acetic acid = 3 part vol.: 1 part vol.): Take 3 part of 100% methanol and 1 part of 100% acetic acid (e.g., 150 mL of methanol and 50 mL of acetic acid for make 200 mL fixative) and mix together well. After mixing, the temperature will be increased according to exothermic reaction (*see* **Notes 5**).
4. 10 mM phosphate buffer (pH 6.8 at 25 °C): Dissolve 136.09 g KH_2PO_4 with 1 L Milli-Q water to make 1 M stock solution A and dissolve 174.18 g K_2HPO_4 with 1 L Milli-Q water to make 1 M stock solution B. Mix 50.3 mL solution A with 49.7 mL solution B, and then make it up to 1 L with Milli-Q water to make 10 mM working solution. If desired, adjust the pH before the final mess up to 1 L (*see* **Note 6).**
5. Giemsa dye solution: Usually suppled from makers as 50 % concentration dye dissolved in 100 % methanol. Dilute with 10 mM phosphate buffer (pH 6.8 at 25 °C) to make 5% (4-6%) solution (freshly prepare just before use) (*see* **Note** 7).
6. Glass slides and cover slips: Glass slides are soaked in 100% methanol until use. Wipe out methanol with lint-free tissue paper.

3 Methods

1. Drug-induced premature chromosome condensation protocol is somehow different between for adherent cells and for suspension cells, although they are basically similar. Most critical point is choice of drugs to induce PCC in these cells. To induce PCC in adherent cells, it is absolutely necessary to use calyculin A. Because okadaic acid does not usually induce PCC in

adherent cells. The reason is still not resolved. In contrast, both okadaic acid and calyculin A are able to induce PCC in suspension cells.

2. Drug-induced premature chromosome condensation (PCC) for adherent cells.

   2.1. Cell culture, premature chromosome condensation and cell harvest.

   Plating the cells in culture dish 1–2 days before the practice in order for the cells to reach 70–80% confluent and growing exponentially at the induction of PCC to achieve a high PCC index number. One 35 mm diameter culture dish of cells is sufficient to obtain substantial numbers of PCCs for analyzing the one data point. Add 1/2000 volume of 100 μM stock solution of calyculin A (i.e., 50 nM final concentration) to the medium, and then incubate for another 30 min at 37 °C. After incubation, cells were gently pipetted with the medium using a 1000 mL micropipette with a blue tip, and cells easily come loose and are suspended in the medium. It requires neither trypsinization nor the use of scraper for detaching cells is usually done following colcemid treatment, because the cells become round shape and attach very loosely on the dishes after calyculin A exposure. If the cells do not become round and do not come loose from the dish, it suggests the possible activity loss of calyculin A. Therefore, a challenge again using a new aliquot calyculin A will be recommended. Transfer the cells suspended in the medium into a centrifuge tube; centrifuge at 250 *g* for 5 min at room temperature, and discard the supernatant by inverting the tube quickly (*see* **Notes 8 and 9**).

   2.2. Chromosome preparation.

   This protocol is for 35 mm dish culture scale. Loosen the cell precipitation on the bottom of tube by quick shaking the tube with swing wrist; add gently 37 °C pre-warmed 1.5 mL 0.075 M KCl to suspend the cells for hypotonic treatment. Incubate at 37 for 20 min. Fix cells by slowly adding the same volume (i.e., 1.5 mL) of chilled Carnoy's fixative to the medium and mix well gently by inverting the tubes a couple of times. Centrifuge at 250 *g* for 5 min, discard the supernatant, loose the cell pellet, add the fixative, and centrifuge, as mentioned above (*see* **Notes 8 and 9).** Repeat a total of three to four times these steps for complete fixation. After fixation, cells were finally suspended in adequate volume of fixative (i.e., 200–300 μl for 35 mm dish culture of cells). Aliquot chromosome suspension (~ 15 μL using a Pipetman using a yellow tip) was dropped on a cleaned glass slide and air-dried. To make sure that the cells are certainly condensed prematurely, observe the chromosome spreads under a microscope in low magnification before next staining. As usual, many chromosome spreads are seen like fireworks in the sky, and 10–20% or more PCC index will be achieved (*see* **Note 9**). Stain the chromosome spreads with 5% Giemsa dye solution, cover by a coverslip, and mount with sealant for observation (*see* **Note 10**). Usually, one sample slide may give a substantial number of PCCs for karyotyping or chromosome aberration analysis. Unstained chromosome specimens will be stored in a desiccated container and will be subjected to several special banding procedures (G-, Q- or R-banding), in situ hybridization including FISH (fluorescent in situ hybridization) or immunostaining or other analysis.

3. Drug-induced PCC for suspension cells

   3.1. Cell culture, premature chromosome condensation and harvest

   The protocol for suspension cells such as peripheral blood lymphocytes is basically the same as for adherent cells. For suspension cells, okadaic acid also works well as calyculin A (*see* **Note 11**).

Prepare the suspension cell culture in an appropriate cell concentration (i.e., 10^5–10^6 cells/ml, depends on cell types used) to maintain exponentially growing at the time of PCC induction. Prior to the PCC induction, it is convenient to transfer the cells in a centrifuge tube. One mL of cell culture with adequate cell density is sufficient to obtain a substantial number of cells showing PCCs for analyzing the one data point. Add 1/2000 volume of calyculin A stock solution (100 μM) to the medium (0.5 μL of calyculin A), process as same as the protocol for attached cells.

   3.2. Chromosome preparation

   After 30 min incubation, harvest the cells by centrifuging and subject to hypotonic treatment, cell fixation, and spreading on the glass slid as the same as adherent cells case (*see* Subheadings 2.1 and 2.2 for detailed protocols described in above sections).

4. For basic and fundamental knowledge about PCC, induction of PCC, specific inhibitors for protein serine/threonine phosphatase as inducer of PCC, mechanism of chromosome condensation etc., (*see* **Notes 12, 13, and 14**).

4 Notes

1. Premature chromosome condensation (PCC) usually means the phenomenon or method, and the condensed interphase chromatin is called as prematurely condensed chromosomes (PCCs). They give the similar abbreviations that are usually used interchangeably.
2. As easily recognized, the drug-induced PCC protocol is a very simple whereby substitute use of calyculin A to colcemid in established chromosome preparation protocol. However, there are some keys for succeed in achieving high PCC index and good-quality PCCs. The belows are the keys that we have experienced through the research study.
3. Calyculin A is now available from many chemical suppliers.
4. It is hard to see and weigh out the lyophilized powder in the shipped vial. Therefore, pour the solvent (ethanol or DMSO) into the vial using a micropipette, gently slant and swirl the vial to dissolve the whole chemicals, and transfer the solution to the microcentrifuge tube (e.g., Eppendorf tube). If dissolved in ethanol, the screwed-lid with a rubber seal-type tube is preferable; otherwise, ethanol will escape in vapor and the concentration of calyculin A will change. Accordingly, DMSO is more preferable for the solvent for calyculin A.
5. It is preferable to prepare just before use; however, the fixative can be stored at −20 °C for a month. After being stored in longer time, esterification will proceed resulting in the production of methyl acetate and water. Consequently, fixative smells some aroma, and, dropped on a glass slide, fixative does not spread well circularly, and water will remain after the fixative evaporated. In this case, the fixative got old, and new fixative should be prepared.
6. Gurr Buffer tablet (one tablet for 1 L solution, pH 6.8, available from Invitrogen, Gibco, ThermoFisher, cat. No. 10582013) or phosphate buffer powder (for 1 L solution, 1/15 M, pH 6.8, equivalent products are available from many makers) are also used well for convenient alternative for phosphate buffer described above (*see* Subheading 3.2, **item 4**).
7. Freshly prepare just at use. After being diluted with phosphate buffer, Giemsa dye soon aggregates and staining activity gets reduced.
8. A round bottom centrifuge tube (e.g., Beckton & Dickinson 5 mL polystyrene centrifuge tube with snap cap) is preferable for chromosome preparation, so the cells are less packed after centrifugation as to avoid cells aggregation (*see* Subheadings 3.1 and 3.2, **steps 1 and 2**).

9. To discard the supernatant, simply invert vertically as quick as possible. Do not slant to discard the supernatant; otherwise, the cell layer will slip away from the tube.
10. Usually, calyculin A solely works efficiently and gives a high PCC index. Combinational use of colcemid with calyculin A will be sometimes effective in obtaining the higher PCC index [11].
11. The tint of Giemsa staining varies as the pH of phosphate buffer shifts; color turns from reddish purple (pH 6.8) to bluish purple (pH 7.2). Which pH of phosphate buffer to use depends on the individual favorite, but reddish purple color seems to be better for observation and taking images under a microscope (*see* Subheadings 3.1 and 3.2, **steps 2 and 2**).
12. Okadaic acid, its salts or derivatives, or other inhibitors such as endothal, cantharidin, or cantharidic acid can be also used as PCC inducers, but these works only in suspension-type cells (e.g., peripheral blood lymphocytes) but not in adherent cells. The reason is still unclear and presumably due to the different cell sensitivities to these chemicals. Microcystin-LR is also a strong inhibitor for protein serine/threonine phosphatase. However, microcystin-LR does not induce PCC in cells, because this chemical does not penetrate in cells. Therefore, the use of calyculin A is recommended for PCC induction both in suspension and adherent types of cell. If okadaic acid is used in suspension cells, 100 nM final concentration rather than 50 nM will be work better (*see* Subheadings 3.1 and 3.2, **steps 1 and 1**).
13. The PCC efficiency is highly dependent on cell types, cell growth, and dose/incubation time of calyculin A. Usually, 50 nM final concentration of calyculin A works very well in many kinds of cells, and sufficient numbers of PCCs will be obtained following 30 min of incubation. However, some cells show the resistance to calyculin A, but others are hypersensitive to calyculin A. In such cases, start the experiment at first trial with 50 nM concentration and for 30 min incubation time, and then increase/decrease the dose and/or the incubation time depending on the cell response and resulting PCC index and degree of chromosome condensation. Keep in mind that the cells should be growing exponentially at the induction of PCC; otherwise, PCC index will drop down. In the case of adherent types of cells, cells should be replated 1–2 day(s) before the experiment to reach 70–80% confluence at the time of PCC induction. Avoid replating the cells at the day of the experiment. The higher the dose and the longer the incubation time, the higher the PCC index, but the resulting PCCs become hyper-condensed. Therefore, the appropriate dose and

incubation time should be taken on balance depending on the experiment design. PCC starts 5 min after the beginning of calyculin exposure, and 10 min of incubation will induce a substantial numbers of PCCs [35, 36, 38] (*see* Subheadings 3.1 and 3.2, **steps 1 and 2**).

14. Calyculin A, okadaic acid, or other phosphatase inhibitors induce PCC predominantly in G2- and S-phase but only few in G1- and G0-phase [28]. Presumably, drug-induced PCC exploits the activation of intracellular MPF by phosphatase inhibitors, but the amount of intracellular maturation-promoting factor (MPF, also known as p34cdc2/cyclin B complex) [49–53] is very low in G1- or G0-phase [28, 54]. In the cases where PCC analysis is requested in G0- or G1-phase of cells, the conventional fusion PCC is still worthwhile to do. Another approach of drug-induced PCC in G0-phase peripheral blood lymphocytes was reported using adenosine triphosphate (ATP) and p34cdc2/cyclin B with okadaic acid [55]. The protocol of fusion PCC or okadaic acid/p34cdc2/cyclin B/ATP complex induced PCC have been described previously elsewhere [2, 31, 55–57].

References

1. Gotoh E, Durante M (2006) Chromosome condensation outside of mitosis: mechanisms and new tools. J Cell Physiol 209(2): 297–304. https://doi.org/10.1002/jcp.20720
2. Johnson RT, Rao PN (1970) Mammalian cell fusion: induction of premature chromosome condensation in interphase nuclei. Nature 226(247):717–722
3. Cornforth MN, Bedford JS (1983) X-ray--induced breakage and rejoining of human interphase chromosomes. Science 222(4628): 1141–1143
4. Gotoh E, Asakawa Y, Kosaka H (1995) Inhibition of protein serine/threonine phophatases directly induces premature chromosome condensation in mammalian somatic cells. Biomed Res 16(1):63–68
5. Rao PN, Johnson RT (1970) Mammalian cell fusion: studies on the regulation of DNA synthesis and mitosis. Nature 225(228):159–164
6. Rao PN, Johnson RT (1971) Mammalian cell fusion. IV Regulation of chromosome formation from interphase nuclei by various chemical compounds. J Cell Physiol 78(2):217–223
7. Sperling K, Rao PN (1974) Mammalian cell fusion. V Replication behaviour of heterochromatin as observed by premature chromosome condensation. Chromosoma 45(2):121–131
8. Hanks SK, Gollin SM, Rao PN, Wray W, Hittelman WN (1983) Cell cycle-specific changes in the ultrastructural organization of prematurely condensed chromosomes. Chromosoma 88(5):333–342
9. Gotoh E, Asakawa Y (1996) Detection and evaluation of chromosomal aberrations induced by high doses of gamma-irradiation using immunogold-silver painting of prematurely condensed chromosomes. Int J Radiat Biol 70(5):517–520
10. Asakawa Y, Gotoh E (1997) A method for detecting sister chromatid exchanges using prematurely condensed chromosomes and immunogold-silver staining. Mutagenesis 12(3):175–177
11. Durante M, Furusawa Y, Gotoh E (1998) A simple method for simultaneous interphase-metaphase chromosome analysis in biodosimetry. Int J Radiat Biol 74(4):457–462
12. Hittelman WN, Pollard M (1984) Visualization of chromatin events associated with repair of ultraviolet light-induced damage by premature chromosome condensation. Carcinogenesis 5(10):1277–1285
13. Pantelias GE, Maillie HD (1984) The use of peripheral blood mononuclear cell prematurely condensed chromosomes for biological dosimetry. Radiat Res 99(1):140–150

14. Hittelman WN (1986) Visualization of chromatin events during DNA excision repair in XP cells: deficiency in localized but not generalized chromatin events. Carcinogenesis 7(12): 1975–1980
15. Pantelias GE (1986) Radiation-induced cytogenetic damage in relation to changes in interphase chromosome conformation. Radiat Res 105(3):341–350
16. Hittelman WN (1990) Direct measurement of chromosome repair by premature chromosome condensation. Prog Clin Biol Res 340B:337–346
17. Suzuki M, Watanabe M, Suzuki K, Nakano K, Matsui K (1992) Heavy ion-induced chromosome breakage studied by premature chromosome condensation (PCC) in Syrian hamster embryo cells. Int J Radiat Biol 62(5):581–586
18. Gotoh E, Tanno Y (2005) Simple biodosimetry method for cases of high-dose radiation exposure using the ratio of the longest/shortest length of Giemsa-stained drug-induced prematurely condensed chromosomes (PCC). Int J Radiat Biol 81(5):379–385
19. Gotoh E, Tanno Y, Takakura K (2005) Simple biodosimetry method for use in cases of high-dose radiation exposure that scores the chromosome number of Giemsa-stained drug-induced prematurely condensed chromosomes (PCC). Int J Radiat Biol 81(1):33–40
20. Gotoh E (2007) Visualizing the dynamics of chromosome structure formation coupled with DNA replication. Chromosoma 116(5): 453–462. https://doi.org/10.1007/s00412-007-0109-5
21. Ono T, Yamashita D, Hirano T (2013) Condensin II initiates sister chromatid resolution during S phase. J Cell Biol 200(4):429–441. https://doi.org/10.1083/jcb.201208008
22. Voisin P (2015) Standards in biological dosimetry: a requirement to perform an appropriate dose assessment. Mutat Res Genet Toxicol Environ Mutagen 793:115–122. https://doi.org/10.1016/j.mrgentox.2015.06.012
23. Lu X, Zhao H, Feng JB, Zhao XT, Chen DQ, Liu QJ (2016) Dose response of multiple parameters for calyculin A-induced premature chromosome condensation in human peripheral blood lymphocytes exposed to high doses of cobalt-60 gamma-rays. Mutat Res Genet Toxicol Environ Mutagen 807:47–54. https://doi.org/10.1016/j.mrgentox.2016.06.010
24. Guerrero-Carbajal C, Romero-Aguilera I, Arceo-Maldonado C, Gonzalez-Mesa JE, Cortina-Ramirez GE, Garcia-Lima O (2019) Dose response of prematurely condensed chromosome rings after gamma irradiation. Int J Radiat Biol 95(5):607–610. https://doi.org/10.1080/09553002.2019.1566677
25. Sun M, Moquet J, Barnard S, Lloyd D, Ainsbury E (2020) A simplified Calyculin A-induced premature chromosome condensation (PCC) protocol for the Biodosimetric analysis of high-dose exposure to gamma radiation. Radiat Res 193(6):560–568. https://doi.org/10.1667/rr15538.1
26. Cornforth MN, Bedford JS (1983) High-resolution measurement of breaks in prematurely condensed chromosomes by differential staining. Chromosoma 88(4):315–318
27. Lau YF, Arrighi FE (1980) Studies of mammalian chromosome replication. I BrdU-induced differential staining patterns in interphase and metaphase chromosomes. Cytogenet Cell Genet 27(2–3):176–183
28. Gotoh E, Durante M (2006) Chromosome condensation outside of mitosis: mechanisms and new tools. J Cell Physiol. 209:297–304. and cover page.
29. Iliakis GE, Pantelias GE (1990) Production and repair of chromosome damage in an X-ray sensitive CHO mutant visualized and analysed in interphase using the technique of premature chromosome condensation. Int J Radiat Biol 57(6):1213–1223
30. Pantelias GE, Iliakis GE, Sambani CD, Politis G (1993) Biological dosimetry of absorbed radiation by C-banding of interphase chromosomes in peripheral blood lymphocytes. Int J Radiat Biol 63(3):349–354
31. Durante M, George K, Wu H, Yang TC (1996) Rejoining and misrejoining of radiation-induced chromatin breaks. I experiments with human lymphocytes. Radiat Res 145(3): 274–280
32. Yamashita K, Yasuda H, Pines J, Yasumoto K, Nishitani H, Ohtsubo M, Hunter T, Sugimura T, Nishimoto T (1990) Okadaic acid, a potent inhibitor of type 1 and type 2A protein phosphatases, activates cdc2/H1 kinase and transiently induces a premature mitosis-like state in BHK21 cells. EMBO J 9(13):4331–4338
33. Schlegel R, Belinsky GS, Harris MO (1990) Premature mitosis induced in mammalian cells by the protein kinase inhibitors 2-aminopurine and 6-dimethylaminopurine. Cell Growth Differ 1(4):171–178
34. Ghosh S, Paweletz N, Schroeter D (1992) Failure of kinetochore development and mitotic spindle formation in okadaic acid-induced premature mitosis in HeLa cells. Exp Cell Res

201(2):535–540. https://doi.org/10.1016/0014-4827(92)90307-t

35. Gotoh E (2007) Visualizing the dynamics of chromosome structure formation coupled with DNA replication. Chromosoma [Epub ahead of printing]
36. Bryant PE, Mozdarani H (2007) A comparison of G2 phase radiation-induced chromatid break kinetics using calyculin-PCC with those obtained using colcemid block. Mutagenesis [Epub ahead of printing]
37. Johnson RT, Gotoh E, Mullinger AM, Ryan AJ, Shiloh Y, Ziv Y, Squires S (1999) Targeting double-strand breaks to replicating DNA identifies a subpathway of DSB repair that is defective in ataxia-telangiectasia cells. Biochem Biophys Res Commun 261(2):317–325
38. Gotoh E, Kawata T, Durante M (1999) Chromatid break rejoining and exchange aberration formation following gamma-ray exposure: analysis in G2 human fibroblasts by chemically induced premature chromosome condensation. Int J Radiat Biol 75(9):1129–1135
39. IAEA (2001) Cytogenetic analysis for radiation dose assesment. A manual. Technical reports series no. 405. International Atomic Energy Agency, Vienna
40. Bezrookove V, Smits R, Moeslein G, Fodde R, Tanke HJ, Raap AK, Darroudi F (2003) Premature chromosome condensation revisited: a novel chemical approach permits efficient cytogenetic analysis of cancers. Genes Chromosomes Cancer 38(2):177–186
41. El Achkar E, Gerbault-Seureau M, Muleris M, Dutrillaux B, Debatisse M (2005) Premature condensation induces breaks at the interface of early and late replicating chromosome bands bearing common fragile sites. Proc Natl Acad Sci U S A 102(50):18069–18074
42. Srebniak MI, Trapp GG, Wawrzkiewicz AK, Kazmierczak W, Wiczkowski AK (2005) The usefulness of calyculin A for cytogenetic prenatal diagnosis. J Histochem Cytochem 53(3): 391–394
43. Mochida A, Gotoh E, Senpuku H, Harada S, Kitamura R, Takahashi T, Yanagi K (2005) Telomere size and telomerase activity in Epstein-Barr virus (EBV)-positive and EBV-negative Burkitt's lymphoma cell lines. Arch Virol 150(10):2139–2150. https://doi.org/10.1007/s00705-005-0557-2
44. Deckbar D, Birraux J, Krempler A, Tchouandong L, Beucher A, Walker S, Stiff T, Jeggo P, Lobrich M (2007) Chromosome breakage after G2 checkpoint release. J Cell Biol 176(6):749–755
45. Kanda T, Kamiya M, Maruo S, Iwakiri D, Takada K (2007) Symmetrical localization of extrachromosomally replicating viral genomes on sister chromatids. J Cell Sci 120(Pt 9): 1529–1539
46. Sun M, Moquet J, Lloyd D, Ainsbury E (2020) A faster and easier biodosimetry method based on calyculin A-induced premature chromosome condensation (PCC) by scoring excess objects. J Radiol Prot 40(3):892–905. https://doi.org/10.1088/1361-6498/aba085
47. Soni A, Murmann-Konda T, Magin S, Iliakis G (2019) A method for the cell-cycle-specific analysis of radiation-induced chromosome aberrations and breaks. Mutat Res 815: 10–19. https://doi.org/10.1016/j.mrfmmm.2019.04.001
48. Okayasu R, Liu C (2019) G1 premature chromosome condensation (PCC) assay. Methods Mol Biol 1984:31–38. https://doi.org/10.1007/978-1-4939-9432-8_4
49. Masui Y (1974) A cytostatic factor in amphibian oocytes: its extraction and partial characterization. J Exp Zool 187(1):141–147
50. Masui Y (2001) From oocyte maturation to the in vitro cell cycle: the history of discoveries of maturation-promoting factor (MPF) and cytostatic factor (CSF). Differentiation 69(1):1–17
51. Dunphy WG, Brizuela L, Beach D, Newport J (1988) The Xenopus cdc2 protein is a component of MPF, a cytoplasmic regulator of mitosis. Cell 54(3):423–431
52. Gautier J, Norbury C, Lohka M, Nurse P, Maller J (1988) Purified maturation-promoting factor contains the product of a Xenopus homolog of the fission yeast cell cycle control gene cdc2+. Cell 54(3):433–439
53. Maller J, Gautier J, Langan TA, Lohka MJ, Shenoy S, Shalloway D, Nurse P (1989) Maturation-promoting factor and the regulation of the cell cycle. J Cell Sci Suppl 12:53–63
54. Doree M, Galas S (1994) The cyclin-dependent protein kinases and the control of cell division. FASEB J 8(14):1114–1121
55. Prasanna PG, Escalada ND, Blakely WF (2000) Induction of premature chromosome condensation by a phosphatase inhibitor and a protein kinase in unstimulated human peripheral blood lymphocytes: a simple and rapid technique to study chromosome aberrations using specific whole-chromosome DNA hybridization probes for biological dosimetry. Mutat Res 466(2):131–141
56. Pantelias GE, Maillie HD (1983) A simple method for premature chromosome

condensation induction in primary human and rodent cells using polyethylene glycol. Somatic Cell Genet 9(5):533–547. https://doi.org/10.1007/bf01574257

57. Prasanna PG, Blakely WF (2005) Premature chromosome condensation in human resting peripheral blood lymphocytes for chromosome aberration analysis using specific whole-chromosome DNA hybridization probes. Methods Mol Biol 291:49–57

Chapter 6

Apoptosis Detection Assays

Ilham Alshiraihi and Takamitsu A. Kato

Abstract

Many apoptosis assays are available since there are many proteins regulated at multiple points and involved in apoptosis signaling cascade. To detect apoptosis accurately, two or more assays should be used since there are many overlapped features between apoptosis and necrosis. There are six major groups of available assays to detect apoptosis: membrane alteration, mitochondrial assays, cytomorphological alterations, DNA fragmentation, detection of caspase, cleaved substrate, inhibitors and regulators, and detection of apoptosis in whole mounts. Among those assay, early apoptosis could be detected through annexin V, which is based on the loss of the cellular membrane integrity. Also, there are many assays that can detect midphase of apoptosis using caspase activation and molecular processing including PARP degradation. Late phase of apoptosis could be detected with DNA fragmentation assays. Combinations of these assays allow us to identify the mechanisms of apoptosis induction after specific stimulus. This chapter will introduce three apoptosis detection assays including annexin assay, DNA/chromatin condensation assays, and TUNEL assay.

Key words Apoptosis, Annexin V, DNA condensation, TUNEL assay

1 Introduction

1.1 Apoptosis and Cell Death

The study of apoptosis began in the eighteenth century when Robert Hook noticed a few cells in the bark of the cork oak and were undergoing physiological death as he interpreted [1]. In the nineteenth century, Carl Vogt noticed and described dying cell in the nervous system when he was studying toad embryos [1]. The French anatomists R. Collin and the German anatomists M. Ernst and A. Glucksmann at the beginning of the twentieth century reported cell death during embryonic development. In 1940, a reasonable demonstration appeared for this type of cell death when Rita Levi-Montalcini noticed that during embryonic development, neuron death (including in the central nerve system) usually occurs [2]. Then, Viktor Hamburger, Levi-Montalcini's colleague, explained that during embryonic development in chicken, approximately half of the motor neuron cells located

Eisuke Gotoh (ed.), *Chromosome Analysis: Methods and Protocols*, Methods in Molecular Biology, vol. 2519,
https://doi.org/10.1007/978-1-0716-2433-3_6,

laterally of the spinal cord will undergo death [3]. In 1972, Kerr, Wyllie, and Currie published the first paper that used the term apoptosis (a-po-toe-sis), which is a Greek term meaning "filling off." In this paper, they described the morphology of the cell death, even though some of the apoptosis components had been detected many years previously [4–6].

The process of apoptosis in mammalian cells has been understood through the examination of programed cell death that happens in nematode Caenorhabditis elegans during the development [7]. Around 131 cells out of 1090 cells, which are generated in this organism to form the adult worm, undergo apoptosis or programed cell death. At specific points of the development process in this worm, these 131 cells die reflecting the system precision. Programed cell death as a mode of apoptosis, which determines genetically the elimination of the cells, has since been recognized and understood. Even though it is important to know that some forms of programed cell death have been recognized, and others have not yet [8–10].

During development and aging, apoptosis normally happens as a homeostatic mechanism to maintain cell population in tissue. Apoptosis also could be triggered as a response to physiological and pathological motivates, such as a defense response to diseases or noxious agents [11]. However, not all cells necessarily undergo apoptosis as reaction to the stimulus. For example, cancer chemotherapy used irradiation and drugs to cause DNA damage in some cells through a p53-dependent pathway. Also hormones, such as corticosteroids, may led to apoptosis in some cells without affecting others [12].

Cells undergoing apoptosis can be determined by many assays. Apoptotic cells' morphology has been detected using light and electron microscope [13]. Cell shrinkage and pyknosis, an irreversible condensation of chromatin in the nucleus of a cell, are signs for early apoptosis and are detected by light microscope [4]. The cellular cytoplasm becomes dense, the organelles become more packed, and cells become smaller when the cell shrinks in early apoptosis process. The most characteristic feature of apoptosis is chromatin condensation that leads to pyknosis. Apoptotic cells are round or oval with dark cytoplasm and chromatin fragments in the nucleus which can be detected by histologic examination using hematoxylin and eosin. Subcellular changes of chromatin condensation phase, such as the dense nuclear peripheral aggregations under the nuclear membrane, can be detected using electron microscope [12]. Then, a process called "budding" occurs when apoptotic bodies form, as a result of blebbing of the plasma membrane, followed by karyorrhexis, a fragmentation of the nucleus and the breakup of the chromatin, and separation of cell fragments. Cellular cytoplasm and packed organelles either with or without a nuclear fragment form apoptotic body. Phagocytosis, through macrophages, neoplastic cells, or parenchymal cells, and degradation within phagolysosomes are the following process to digest the apoptotic bodies [12].

Apoptosis mechanisms involve an energy-dependent cascade of molecular events and are very complicated. There are many apoptotic pathways and two main pathways that are indicated in research: the intrinsic or mitochondrial pathway and the external, or death, receptor pathway. The intrinsic pathway is led by mitochondria that stimulate apoptosis by releasing caspase-activating protein into the cytosol. The external pathway depends on tumor necrosis factor (TNF). Apoptosis in this case is induced through the binding of ligand at the cell surface which triggers caspase cascade mechanism [14]. There is another pathway depending on perforin-granzyme-dependent killing of the cells and T-cell-mediated cytotoxicity. Granzyme B or granzyme A mediate apoptosis in the perforin/granzyme pathway. These pathways are connected on the execution pathway, and molecules in one pathway could influence the other one, as was proved by previous studies. Execution pathway is started through cleavage of caspase-3 and leads to DNA fragmentation, degradation of nuclear proteins, and cytoskeleton and formation of apoptotic bodies [15].

1.2 Annexin V

One of the most important signs of early apoptosis is the externalization of the phosphatidylserine residues from the inner to the outer cell membrane, resulting in easy detection by annexin [16]. Annexin V, usually conjugated with a wide variety of dyes, bind with the phosphatidylserine and can be visualized by fluorescent microscopy. This marker is very sensitive and able to detect the activity of initiator caspases; however, necrotic cells membranes could be labeled as well. Therefore, determining phosphatidylserine-positive cells is very important to demonstrate the membrane integrity. Propidium iodide and trypan blue are a few of specific membrane-impermeant nucleic acid dyes that can be used to examine the loss of membrane integrity as a pathognomonic sign of necrotic cell death. The exclusion of these dyes is a sign of the membrane integrity of apoptotic cells. However, the externalization of phosphatidylserine to the outer cell membrane will also allow nucleic acid dyes to transport into the cell. This dye-uptake bioassay is sensitive because it can detect a single apoptotic cell without labeling necrotic cells [12].

1.3 DNA Condensation

Detection of chromatin condensation as a sign of apoptosis using DAPI staining approach is one of the convenient methods for analysis of late apoptosis. This method is suitable to distinguish healthy cells and cells that undergo apoptosis. DAPI is one of the minor groove-binding dyes that are highly fluorescent once it is bound to DNA. Therefore, the principle of this approach is that DAPI stains nucellar-compacted chromatin of apoptotic cells in higher amount than healthy cells. Because DAPI stain is less permeable than other staining as Hoechst, it is typically used to stain fixed cells at relatively high concentrations (1 μg/mL). This approach provides rapid checking of healthy cells and can be used for a quick screening for agents that could cause apoptosis.

1.4 TUNEL Assay

The TUNEL (terminal dUTP nick-end labeling) assay is one of the methods to detect apoptosis, and it is based on detection of endonuclease cleavage products using enzymatically end labeling the DNA strand breaks [17]. In this assay, addition of labeled UTP to the 3′-end of the DNA fragments is mediated by terminal transferase. Then, a variety of probes can be used to label the dUTP and detected using flow cytometry, light microscopy, or fluorescence microscopy. The advantages of this assay are that it is fast and can be completed within 3 h, and it is also very sensitive and can be used to detect a single cell by fluorescence microscopy or many cells by flow cytometry. On the other hand, the disadvantages of this assay are that it is expensive, and sensitivity to detect DNA fragmentation may vary among laboratories with minor protocol modifications. This assay also should be done with another apoptotic assay because it may undergo false-positives from cells that undergo DNA repair and cells that undergo necrosis [12].

2 Materials and Equipment

2.1 Materials

1. Cells are cultured in appropriate cell culture media for individual cell line such as MEM (Minimum Essential Media), alpha-MEM, DMEM (Dulbecco's Modified Essential Medium), and RPMI (Roswell Park Memorial Institute 1640) supplemented with fetal bovine serum (FBS) and antibiotic. Prior lot testing is required for FBS. Some lots increase background frequency of SCE formation. Heat inactivation of complement at 56 °C for 20 min is required for FBS. For antibiotics such as Antibiotic-Antimycotic (100x), penicillin-streptomycin solution (5000 U/ml) should be added.
2. Phosphate-buffered saline (−) (PBS).
3. Trypsin-EDTA (0.05% or 0.25% Trypsin/EDTA-4Na in PBS).
4. 75 mM hypotonic potassium chloride (KCl) solution.
5. Carnoy's fixative solution (methanol and acetic acid = 3 vol: 1 vol), prepare at time of use.
6. Propidium iodide (PI) 1 mg/ml in PBS.
7. Annexin V APC and annexin-binding buffer 10X (100 mM HEPES pH 7.4, 50 mM KCI, 1.5 M NaCl, 18 mM $CaCl_2$, 10 mM $MgCl_2$).

8. Mouse monoclonal anti-BrdU antibody (such as clone BU1/75 (ICR1), BU-1 or B44).
9. Alexa488-conjugated goat anti-mouse IgG antibody.
10. DAPI in SlowFade.
11. 4% (w/v) paraformaldehyde solution in PBS.
12. 0.2% Triton X-100 in PBS solution.
13. TUNEL 5x reaction buffer: 1 M potassium cacodylate, 125 mM HCl, pH 6.6, Br-dUTP stock solution: 2 mM Br-dUTP in 50 mM Tris HCl pH 7.5, 10 mM $CoCl_2$.
14. Terminal deoxynuleotidyl transderase (TdT).
15. Apoptotic agents: hydrogen peroxide, UV-C, etc.

2.2 Equipment

1. Cell culture facility including biosafety cabinet or clean bench, 37 °C CO_2 controlled incubator, and water bath.
2. Centrifuge, which can rotate 15 ml conical tubes.
3. Aspirator to remove supernatant and trap waste.
4. Flow cytometer and software to analyze data such as Cytek 4-laser Aurora instrument connected with a computer that has SpectroFlo software.
5. Fluorescent microscope with appropriate filter sets equipped with CCD camera and software for fluorescent staining with DNA condensation and TUNEL staining.
6. Slide glass, coverslip.

3 Methods

3.1 Annexin PI Method

1. Culture the cells in appropriate and supplemented media such as DMEM/F12, DMEM, RPMI, and MEM. Plate 3×10^5 cells per mL in 6-well plate.
2. Induce apoptosis in cells using apoptotic agents such as hydrogen peroxide and cisplatin (chemical compound at gradual concentrations 10–200 μM) at respective time points (24–72 h).
3. Prepare the following control samples to use as references for the software:

   (1) Unstained sample, (2) annexin only sample with apoptosis induced by heat shock. (3) PI only with apoptosis induced by heat shock. (4) Annexin V and PI with apoptosis induced by heat shock. (5) Annexin V and PI without apoptosis induced by heat shock. (6) Annexin V and PI as a control group or with apoptosis induced by desired method as a treatment group.

4. Trypsinize cells and transfer to 15 mL tube.
5. Centrifuge the cells at 1000 rpm for 3 min. Remove supernatant and add 2 ml of cold PBS.
6. Induce apoptosis by heat shock method in annexin only sample and PI only sample and annexin V and PI sample using water bath at 60 °C for 5 min.
7. Recentrifuge all samples at 1000 rpm for 3 min and discard the supernatant.
8. Using distilled water, dilute 10X annexin-binding buffer to 1X (1 ml 10X binding buffer in 9 ml distilled water). Wash the cells with 2 ml of 1X binding buffer, recentrifuge the cells at 1000 rpm for 3 min, and discard the supernatant. Repeat this step twice.
9. Resuspend the cells in 100 μl of 1x annexin-binding buffer.
10. Add 2.5 μl of fluorochrome-conjugated annexin V to all samples except unstained sample and PI only sample (*see* **Note 1**).
11. Incubate the samples for 10 min in the dark at room temperature.
12. Wash the cells with 500 μl of annexin-binding buffer, recentrifuge the cells at 1000 rpm for 3 min, and discard the supernatant. Repeat this step twice.
13. Resuspend the cells in 500 μl of Annexin binding buffer.
14. Prepare a 100 μg/mL final working solution of PI by adding 45 μL of 1X Annexin binding buffer to 5 μL of 1 mg/mL PI stock solution (*see* **Note 2**).
15. Add 2.5ul of 100 μg/ml PI to all samples except unstained sample and annexin only sample.
16. Incubate the samples 10 min in the dark at room temperature. Mix gently and keep the samples on ice till analysis. As soon as possible, analyze the samples by flow cytometry using flow cytometer, measuring the fluorescence emission at 650/660 nm.
17. Measure the unstained sample, annexin only with heat shock and then PI only with heat shock. Check the emission peaks and set the gates.
18. The population should separate into four groups: live cells will show only a very low level of fluorescence, early apoptotic will be positive for only annexin, late apoptosis will be double-positive for both annexin and PI, and necrosis will be positive for PI only (Fig. 1) (*see* **Note 3**).

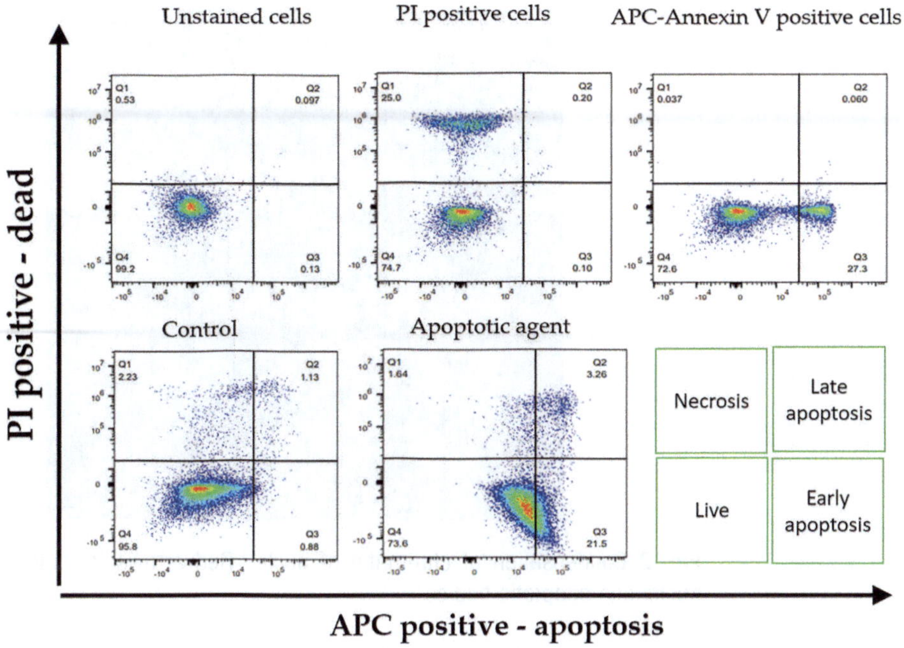

Fig. 1 Cell apoptosis was assessed using APC annexin V/PI and flow cytometry. Unstained cells, PI-positive cells, APC annexin V-positive cells, control (untreated) cells, and apoptosis after 48 h of apoptotic agent

3.2 Slide Preparation for DNA Condensation Analysis and TUNEL Stain

1. $3x10^5$ cells per mL of cells should be cultured in 15 mL conical tube (*see* **Note 4**).

   Appropriate amount of drug or UV and IR should be exposed or irradiated. H_2O_2 can be positive control for apoptosis induction (*see* **Note 5**).
2. Incubate Cells to Induce Apoptosis; 24 h Would Be Good Initial Time Point

   Some agents and cells need 48 h to induce apoptosis.
3. 3.2.3. After incubation, cells should be collected with centrifugation (300 g, 5 min).
4. 3.2.4. Supernatant is removed and pellet should be resuspended in 3 ml of 75 mM KCl solution. Cell pellet should be disturbed with gentle tapping of tubes.
5. 3.2.5. 1 ml of methanol: acetic acid (3:1) fixative solution should be added to this.
6. Cells should be collected with centrifugation (300 g, 5 min).
7. Cell pellet should be resuspended with 1 mL of fixative solution. Repeat centrifugation and resuspend with fixative two more times.
8. After final centrifugation, cell suspension should be dropped on glass slides. Glass slides should be precleaned and immersed in iced water.
9. Slides should be air-dried.

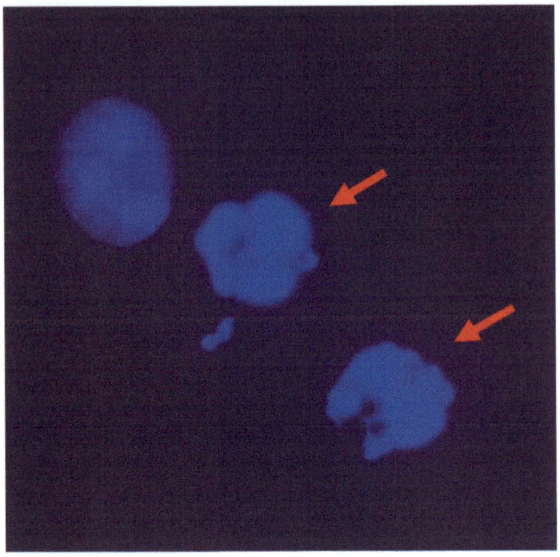

Fig. 2 Condensation of chromatin of nuclei. Red arrows indicate condensed nuclei and apoptotic bodies

3.3 DNA Condensation Analysis

1. (From Subheading 3.2, **step 9**). Add 1 drop of DAPI solution in SlowFade and place coverslip on it.
2. Slides may be stored in fridge up to 1 month with good quality, but analysis should be done within a week for better resolution.
3. In fluorescence microscope, oil immersion 40x, 60x, or 100x objective lens should be used for researchers' preference. DAPI signal (blue) will show nuclei. During apoptosis, compaction of nuclear chromatin occurs. Nuclei are condensed and form apoptotic bodies (Fig. 2). Apoptotic cells show brighter DAPI signal compared to non-apoptotic cells.
4. Minimum 100 cells should be analyzed. The fraction of apoptotic cells should be obtained (*see* **Note 6**).

3.4 TUNEL Assay

1. From Subheading 3.2, **step 9**, put slides into 4% paraformaldehyde in PBS solution in Coplin jar for 15 min.
2. Rinse with PBS.
3. Put slides into 0.2% Triton X-100 in PBS in Coplin jar for 10 min.
4. Rinse with tap water.
5. TUNEL reaction buffer should be placed on the slide and incubate for 4 h at 37 °C. For 50 μL, mix 10 μL of TdT 5x buffer, 2 μL of Br-dUTP stock solution (final 80 μM), 0.5 μL of TdT (final 12.5 units), 5 μL of $CoCl_2$ (final 2 mM), and 33.5 μL of distilled water.
6. Wash slides with PBS.

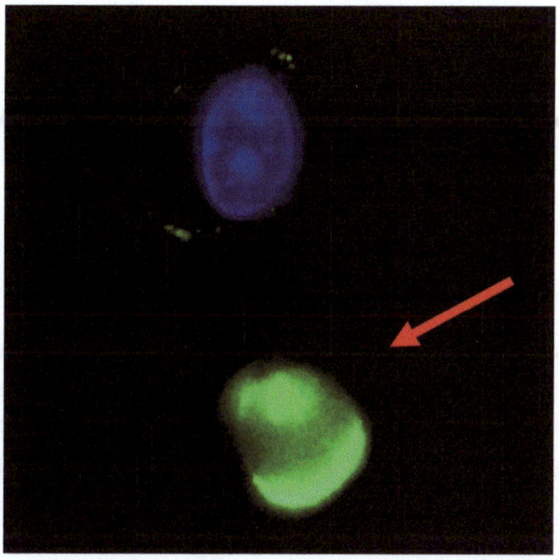

Fig. 3 TUNEL assay. TUNEL stain-positive apoptotic cells and non-apoptotic cells. TUNEL stain label DNA damage with BrdUTP by terminal deoxynucleotidyl transferase (TdT). A red arrow indicates anti-BrdU antibody with Alexa 488 detects the incorporated BrdUTP

7. Add mouse monoclonal BrdU antibody and incubate for 1 h at 37 °C. 1:100–500 dilution in 10% goat serum in PBS.
8. Wash slides with PBS.
9. Add 1:500 diluted Alexa 488 Fluor-conjugated goat anti-mouse antibody and incubate for 1 h at 37 °C.
10. Wash with PBS. Slides should be mounted with coverslip with DAPI.
11. Observe under fluorescent microscope (Fig. 3). At least 40x, preferably 100x objective lens should be used with oil. At least 100 cells should be analyzed. Green signal-positive cells are apoptotic cells.

4 Note

1. During annexin V experiment, it is very important to avoid buffers containing EDTA or other calcium chelators because of the calcium dependence on the annexin V:PS interaction.
2. PI is a sticky dye, so it is very important to clean the instrument probe between the samples (especially after PI only sample and before annexin only sample) to avoid cross-contamination.
3. Nonspecific binding of annexin V to PS inside the cells could happen when the plasma membrane undergoes damage and loses its normal integrity. However, annexin V can be used to detect apoptosis in intact cell membrane. Nonspecific binding

result also could appear in case of use a fluorochrome-conjugated annexin V, which has an emission wavelength overlaps with a fluorescence intercalating agent.

4. Apoptosis analysis should be done with cells easily going to apoptosis such as TK6, Molt4, lymphoblast cells. Fibroblast cells rarely go to apoptosis.
5. H_2O_2 concentration has to be optimized because very low yield of apoptosis induction in low concentration and necrosis may be dominant in high concentration of H_2O_2 treatment. 5 to 50 μM would be sufficient. 200 μM and above may cause more necrosis and less apoptosis. Each cell line should be optimized for the concentration of positive control.
6. This analysis is very quick and easy protocol. Therefore, it would be suitable for screening drugs and determining suitable concentrations and timing of apoptosis analysis.

Apoptotic body formation would be latter part of apoptosis. If researchers need to determine the timing of analysis, early apoptosis markers, such as membrane-associated markers, may occur much earlier than this apoptotic body formation. Apoptotic body formation is an evidence of apoptosis, but other apoptotic assays such as PARP cleavage, annexin V/PI analysis, and caspase 3 activation would determine more precise timing and molecular pathways of apoptosis.

References

1. Vaux DL, Korsmeyer SJ (1999) Cell death in development. Cell 96(2):245–254
2. Oppenheim R, von Bartheld C (1999) In: Zigmond MJ, Bloom FE, Landis SC, Roberts JL, Squire LR (eds) Fundamental neuroscience. Academic Press, San Diego, pp 581–609
3. Jessell TM, Kandel ER, Schwartz JH (2001) Principios de Neurociencia. KAN 577.25
4. Kerr J, Wyllie A Currie Ar (1972) A basic biological phenomenon with wideranging implications in tissue kinetics. Br J Cancer 26: 239
5. Kerr JF (2002) History of the events leading to the formulation of the apoptosis concept. Toxicology 181:471–474
6. Paweletz N (2001) Walther Flemming: pioneer of mitosis research. Nat Rev Mol Cell Biol 2(1):72–75
7. Horvitz HR (1999) Genetic control of programmed cell death in the nematode Caenorhabditis elegans. Cancer Res 59(7 Suppl):1701s–1706s
8. Debnath J, Baehrecke EH, Kroemer G (2005) Does autophagy contribute to cell death? Autophagy 1(2):66–74
9. Formigli L, Papucci L, Tani A, Schiavone N, Tempestini A, Orlandini G, Capaccioli S, Zecchi Orlandini S (2000) Aponecrosis: morphological and biochemical exploration of a syncretic process of cell death sharing apoptosis and necrosis. J Cell Physiol 182(1):41–49
10. Sperandio S, de Belle I, Bredesen DE (2000) An alternative, nonapoptotic form of programmed cell death. Proc Natl Acad Sci 97(26):14376–14381
11. Norbury CJ, Hickson ID (2001) Cellular responses to DNA damage. Annu Rev Pharmacol Toxicol 41(1):367–401
12. Elmore S (2007) Apoptosis: a review of programmed cell death. Toxicol Pathol 35(4): 495–516
13. Häcker G (2000) The morphology of apoptosis. Cell Tissue Res 301(1):5–17. https://doi.org/10.1007/s004410000193

14. Igney FH, Krammer PH (2002) Death and anti-death: tumour resistance to apoptosis. Nat Rev Cancer 2(4):277–288. https://doi.org/10.1038/nrc776
15. Martinvalet D, Zhu P, Lieberman J (2005) Granzyme a induces caspase-independent mitochondrial damage, a required first step for apoptosis. Immunity 22(3):355–370. https://doi.org/10.1016/j.immuni.2005.02.004
16. Bossy-Wetzel E, Green DR (2000) Detection of apoptosis by annexin V labeling. Methods Enzymol 322:15–18
17. Kressel M, Groscurth P (1994) Distinction of apoptotic and necrotic cell death by in situ labelling of fragmented DNA. Cell Tissue Res 278(3):549–556

Chapter 7

Alkaline Comet Assay to Detect DNA Damage

Kade D. Walsh and Takamitsu A. Kato

Abstract

The comet assay is an effective method for identifying DNA breaks and alkali-labile sites induced by genotoxins. Performed as a single-cell electrophoresis, this assay is especially simplistic, and the results are easily reproducible. DNA breakage can be quantitatively assessed by the induced comet tail regions, which can be measured using a variety of comet software. This protocol will finish within approximately two hours with adequate preparation, and digitized images can be taken using a confocal or standard fluorescence microscopes after staining the cell nucleus with a DNA dye.

Key words Comet assay, Single-cell electrophoresis, Alkaline lysis

1 Introduction

The comet assay was first discovered by Ostling and Johanson in 1984 in order to measure single-stranded breaks resulting in the relaxation of supercoils [1]. This protocol was modified slightly by Singh and colleagues, who utilized alkaline conditions [2]. These alkaline conditions have become the preferable approach to this assay, as this removes the tertiary and quaternary structure of the DNA, as well as partially disrupting the secondary structure to ultimately increase the sensitivity [3]. This protocol can be used to assess DNA damages resulting from induced environmental stressors, such as genotoxins, UV light, and ionizing radiation [4–6]. The primary aim of the assay was to combine the agarose electrophoresis methodology and fluorescence microscopy in order to observe and quantitatively measure the DNA breakage at the level of single cells. By implementing this combination, there was a wide array of advantages over the other techniques common to that period. Advantages include that the assay allows for data collection at the individual cellular level; fewer cells per individual sample are sufficient for statistically significant data, particularly high assay sensitivity [7], and the ubiquitous nature of the assay;

Eisuke Gotoh (ed.), *Chromosome Analysis: Methods and Protocols*, Methods in Molecular Biology, vol. 2519, https://doi.org/10.1007/978-1-0716-2433-3_7,

and it can be performed on a wide variety of species of eukaryotic cell populations [8, 9]. Despite the numerous advantages, there are still a few flaws within this protocol. Perhaps the most inconvenient of them all is that the small sample sizes (50 cells) may be inadequate for samples with high heterogeneity in DNA damage [10]. Besides this downfall, other imperative implications include the requirement of viable cell suspensions and the lack of specificity pertaining to DNA fragment sizes. As it may be assumed, cell samples must also assuredly be viable and not apoptotic [11]. Failure to regard this limitation could result in false-positive DNA damages and greatly exaggerated tail moments. Additionally, this assay is unqualified in providing specific DNA fragment sizes, but the tail moment is used rather to associate a ratio of damaged DNA to undamaged DNA. Despite the limitations, the comet assay has continued to participate in its role as an effective tool to observe DNA breaks. Its simplicity and low cost allow for rapid results and data analysis, and the available measurement software can be easily implemented.

Although we considered the common alkaline method that is most effective for our experimental purposes, it does contain some flaws and may not necessarily always be the best choice. Unsurprisingly, the alkaline protocol is best suited for detection of single strand breaks and alkaline labile sites, a prime approach in examining genotoxic hazards according to the OECD guideline [12]. As observed in a survey conducted within the International Comet Assay Workshops of 2015, >70% of participants claimed to have used the alkaline comet assay regularly over many of the other common protocols [12]. Despite its popularity, there are some significant limitations that can be resolved using other experimental approaches. The comet-FISH protocol, for example, grants the experimenter the ability to study gene-specific DNA damage and repair [12]. This method is often done using fluorescent double-stranded probes, oligonucleotide probes, padlock probes, or strand-specific probes [13]. Each probe type has its limitations, including a lack of gene specificity, high number copy sequences, numerous steps requiring high optimization, and substantial time requirements, respectively [13]. However, there are many tremendous advantages to these methods, including an overall much higher sensitivity that enables detection of specific genomic regions of interest and genotoxicity within ranges that are normally undetectable [13]. Despite these advantages, the alkaline comet assay protocol is still considered the standard for the assessment of genotoxic agents. Within the alkaline comet assay, more specific DNA lesions can also be detected by the addition of the DNA endonuclease formamidopyrimidine DNA glycosylase enzyme (FPG). This addition becomes necessary when interested in genotoxicities at the level of DNA adducts or base modifications such as those commonly attributed to purine oxidation [14]. FPG is not specific to oxidized purines but is recommended for assays using genotoxic compounds that do not directly induce single-stranded breaks [14]. Although the FPG and alkaline comet assay combination is still a novel approach, the inclusion of FPG has demonstrated significant effects on the overall sensitivity of the comet assay, making it a valuable reagent that should be taken under consideration.

This chapter introduces the basic alkaline comet assay protocol with freely available software to analyze the comet tail moments. There are several pitfalls for the protocol, all of which we will denote.

2 Materials and Equipment

2.1 Materials

1. Appropriate cell culture media such as MEM, αMEM, DMEM, or RPMI.
2. Fetal bovine serum (FBS), prior lot testing is required. Heat inactivation at 56 °C for 20 min is required.
3. Phosphate-buffered saline (PBS).
4. Antibiotic solutions such as Antibiotic-Antimycotic (100x), penicillin-streptomycin solution (5000 U/mL).
5. Trypsin-EDTA (0.05% or 0.25% Trypsin/EDTA-4Na in PBS).
6. 1% low melting temperature (LMT) agarose.
7. 1% standard agarose gel.
8. Commercially available glass etching cream.
9. Lysis buffer: 2.5 M NaCl, 100 mM EDTA, 10 mM Tris-base, pH 10 by NaOH, 0.1% Triton X-100.
10. Alkaline running buffer: 0.3 M NaOH, 1 mM EDTA pH 13.0.
11. Neutralization buffer: 0.4 M Tris-HCl pH 7.4.
12. 50 μg/ml propidium iodide (PI) solution for counter staining nuclei.

2.2 Equipment

1. Cell culture facility including biosafety cabinet or clean bench, 37 °C CO_2 controlled incubator.
2. Centrifuge, which can rotate 15 ml tubes. No need for temperature control or a brake.
3. Aspirator to remove supernatant and trap waste.
4. Microscope equipped with CCD camera and imaging software.
5. Slide glass and coverslips.
6. Standard horizontal electrophoresis apparatus with power supply.
7. Warm water bath setup for LMT agarose.
8. Compartments for mixing LMT agarose with cell suspensions such as a microtiter plate.

3 Methods

3.1 Slide Preparation Pre-Etching and Coating

1. This protocol most closely follows the alkaline approach to the comet assay. This enables detection of alkaline-labile sites and single-stranded breaks in addition to double-strand breaks. Pre-etching the glass slides with glass etching cream is recommended prior to handling the cell samples.
2. A uniform layer of 100 μL etching cream can be effectively spread on the glass slide using a micropipette tip or spatula. After allowing the etching cream to solidify on the slides for 30 minutes, the cream must be washed off with tap water (*see* **Note 1**).
3. After the slides were etched, a basal layer of agarose should be placed onto a pre-etched slide so that 500 μL of the 1% standard agarose gel, pre-boiled by a microwave, was added to each slide using a micropipette, ensuring that a thin coat covered the entirety of the slides.
4. Slides with a basal layer of agarose should be dried in 80 °C for about 1 hour. Slides with a basal layer of agarose can be stored in room temperature.
5. Cell samples must be trypsinized and centrifuged at 300 g for 5 min and prepared at 200,000 cells/100 μL in PBS (*see* **Note 2**).
6. While the 5 min centrifugation is taking place, the 1% low-melting temperature (LMT) agarose should be microwaved and prepared for the upcoming slide adherence step. Following melting the LMT agarose, it is also recommended that the agarose be kept in a warm bath at 40 °C to avoid rapid re-solidification.
7. The 100 μL cell suspension must be flicked or vortexed. Now that all of the materials are ready to be transferred and combined on the glass slides, maintaining all materials in close proximity is imperative for the next steps, as agarose solidification can occur rather quickly.
8. Cell suspensions should be mixed with 200 μL of the 1% LMT agarose. This can be effectively done by filling a microtiter plate with each of the respective 100 μL cell suspensions in PBS, followed by individually adding and mixing the 200 μL of the 1% LMT agarose with the cells using a p1000 pipette to accommodate the high viscosity.
9. The samples should then quickly be transferred to their respective slides and covered with a coverslip immediately to avoid agarose re-solidification. In these steps the pipette tips must be discarded after each transfer.

10. Slides must be left to settle for approximately 30 min at 4 °C. Coverslips should then be removed in a manner that avoids unnecessary agarose smearing or disintegration. The following buffer and electrophoresis steps can then be performed. For all buffer incubation steps, we used a Coplin jar to allow all of the slide surface area to be covered in buffer solution.

3.2 Alkaline Lysis and Electrophoresis

1. Slides must be placed in pre-chilled alkaline lysis buffer for at least 1 h at approximately 4 °C. Although 1 h may be sufficient, an overnight incubation may be more effective in allowing DNA break points to unwind.
2. Slides must be placed in the alkaline running buffer solution for 20 min prior to electrophoresis.
3. During this 20 min time slot, it is recommended that the electrophoresis apparatus be prepared with pre-chilled alkaline buffer so that the middle slide platform is completely covered. After 20 min, slides are placed in the middle platform of the electrophoresis apparatus and run for 20 min at 1 V/cm (15 V) (*see* **Note 3**).
4. Once the electrophoresis has been completed, the slides must very carefully be transferred to the neutralization buffer.
5. Slides can then be gently rinsed with tap water in a Couplin jar and allowed to air-dry at room temperature for approximately 10 min.
6. Slides can then be dyed using PI solution or any other similar DNA intercalating counterstain. A coverslip was placed over the slides to avoid drying.
7. The samples are now ready for fluorescence imaging with fluorescence microscope with appropriate filters. Samples were imaged with red filter at 20x magnification or higher magnification for PI staining.
8. Successful images should appear like Fig. 1. At least 50-cell images should be taken for each sample to analyze DNA damage.

3.3 Data Analysis

1. There is a variety of software that can be utilized for comet assay image analysis. In our experiments, we utilized OpenComet (http://cometbio.org). However, other available software includes Komet software, Comet Assay IV, the Imstar Pathfinder, and Comet-FISH methods. Within OpenComet, image analysis measures a variety of physical comet characteristics including: comet area, comet intensity, comet length, comet DNA, head area, head intensity, head length, head DNA, head DNA percent, tail area, tail length, tail DNA, tail DNA percent, olive moment, and tail moment. For the sake of our experiments, we were only interested in the tail moment measurements for each cell and averaged them for each sample.

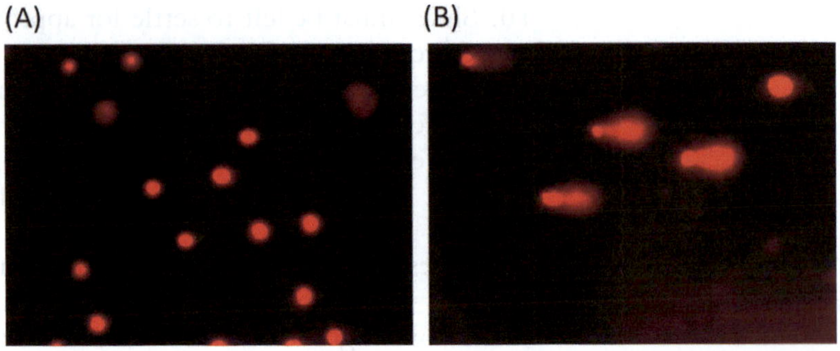

Fig. 1 Successful single-cell electrophoresis comet images. Both the nuclei and comet tail moments can be clearly observed, and resolution is high enough to be effectively measured via a comet software. (**a**) Representation of negative control. (**b**) Representation of DNA damage with corresponding comet tail regions

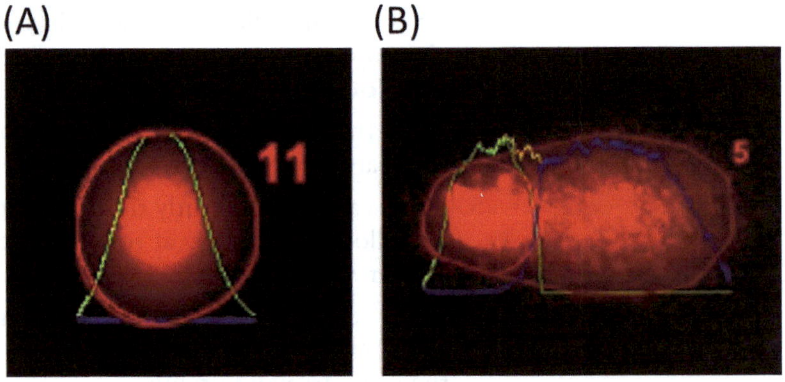

Fig. 2 Comet assay analysis with their associated OpenComet measurements shown below each image, respectively. Green indicates the comet head region and blue indicates the tail region. (**a**) Representation of negative control. (**b**) Representation of DNA damage with corresponding comet tail regions

2. For OpenComet, comet images should be taken like Fig. 1 and stored in a unique folder in a computer to make the following steps more efficient.
3. Within the ImageJ software, choose the image through the "Input files" selection in the OpenComet plugin and select the desired image.
4. Use the "output files" selection to place the image in a desired folder and click "run" to create the image with measurements (Fig. 2).
5. The images obtained should include a variety of cell measurements in an Excel file, but the one that we are most interested in is the "tail moment" column. These data can be taken for each sample to describe the degree of DNA damage present in the sample (*see* **Note 4**).

4 Notes

1. The treatment period with glass etching cream depends on the cream selection, but we used 30 min.
2. In order to assess an effective comet analysis, optimization of cell concentration is necessary. It is crucial that in these steps the viability of the cell suspension is thoroughly checked, and a high density of cells, approximately 200,000 cells/ 100 μL, is optimum.
3. Alkaline buffer is an irritant, and handling slides with gloves and tweezers is highly recommended during lysis and electrophoresis.
4. Although the OpenComet software is effective in measuring every cell in the image, it may not be perfect. It is important that the experimenter check each image with the measurements to make sure that measurements were not made using multiple cells that may be slightly overlapping or cells that may be abnormally shaped. If this occurs, the Excel sheet will likely mark these cells as "outlier," but oftentimes the experimenter may find some simply by observing the measured image. As it implies, these measurements must be excluded from the data collected, as they may drastically and falsely skew the experimental results.

References

1. Ostling O, Johanson KJ (1984) Microelectrophoretic study of radiation-induced DNA damages in individual mammalian cells. Biochem Biophys Res Commun 123(1): 291–298. https://doi.org/10.1016/0006-291x(84)90411-x
2. Singh NP, McCoy MT, Tice RR, Schneider EL (1988) A simple technique for quantitation of low levels of DNA damage in individual cells. Exp Cell Res 175(1):184–191. https://doi.org/10.1016/0014-4827(88)90265-0
3. Singh NP (2016) The comet assay: reflections on its development, evolution and applications. Mutat Res Rev Mutat Res 767:23–30. https://doi.org/10.1016/j.mrrev.2015.05.004
4. Alapetite C, Wachter T, Sage E, Moustacchi E (1996) Use of the alkaline comet assay to detect DNA repair deficiencies in human fibroblasts exposed to UVC, UVB, UVA and gamma-rays. Int J Radiat Biol 69(3): 359–369. https://doi.org/10.1080/095530096145922
5. Moller P (2005) Genotoxicity of environmental agents assessed by the alkaline comet assay. Basic Clin Pharmacol Toxicol 96(Suppl 1): 1–42
6. Bolognesi C, Cirillo S, Chipman JK (2019) Comet assay in ecogenotoxicology: applications in Mytilus sp. Mutat Res 842:50–59. https://doi.org/10.1016/j.mrgentox.2019.05.004
7. Tice RR, Agurell E, Anderson D, Burlinson B, Hartmann A, Kobayashi H, Miyamae Y, Rojas E, Ryu JC, Sasaki YF (2000) Single cell gel/comet assay: guidelines for in vitro and in vivo genetic toxicology testing. Environ Mol Mutagen 35(3):206–221. https://doi.org/10.1002/(sici)1098-2280(2000)35:3<206::aid-em8>3.0.co;2-j
8. Salagovic J, Gilles J, Verschaeve L, Kalina I (1996) The comet assay for the detection of genotoxic damage in the earthworms: a promising tool for assessing the biological hazards of polluted sites. Folia Biol (Praha) 42(1–2):17–21
9. Gajski G, Zegura B, Ladeira C, Pourrut B, Del Bo C, Novak M, Sramkova M, Milic M, Gutzkow KB, Costa S, Dusinska M, Brunborg G,

Collins A (2019) The comet assay in animal models: from bugs to whales - (part 1 invertebrates). Mutat Res 779:82–113. https://doi.org/10.1016/j.mrrev.2019.02.003

10. Olive PL, Banath JP (2006) The comet assay: a method to measure DNA damage in individual cells. Nat Protoc 1(1):23–29. https://doi.org/10.1038/nprot.2006.5
11. Choucroun P, Gillet D, Dorange G, Sawicki B, Dewitte JD (2001) Comet assay and early apoptosis. Mutat Res 478(1–2):89–96. https://doi.org/10.1016/s0027-5107(01)00123-3
12. Koppen G, Azqueta A, Pourrut B, Brunborg G, Collins AR, Langie SAS (2017) The next three decades of the comet assay: a report of the 11th international comet assay workshop. Mutagenesis 32(3):397–408. https://doi.org/10.1093/mutage/gex002
13. Spivak G (2015) New developments in comet-FISH. Mutagenesis 30(1):5–9. https://doi.org/10.1093/mutage/geu036
14. Azqueta A, Arbillaga L, Lopez de Cerain A, Collins A (2013) Enhancing the sensitivity of the comet assay as a genotoxicity test, by combining it with bacterial repair enzyme FPG. Mutagenesis 28(3):271–277. https://doi.org/10.1093/mutage/get002

Chapter 8

Analysis for Sister Chromatid Exchange

Takamitsu A. Kato

Abstract

Sister chromatid exchange (SCE) is the exchange event of genetic material between two identical sister chromatid. Elevation of SCE frequency is considered as a result of replication stress from genetic defects, ROS stress, and genomic damages. SCE staining needs extra processes compared to regular Giemsa staining. Usually two rounds of cell cycle progress are required to observe SCE under microscope. SCE can be visualized with the fluorescence plus Giemsa (FPG) staining method or fluorescence staining methods with immunocytochemistry to BrdU or Click reaction to EdU which provide more clear images of SCE. This chapter will provide the detailed method for the SCE staining and measurement for the traditional FPG staining, BrdU monoclonal antibody staining method, and newly developed EdU Click reaction staining method.

Key words Sister chromatid exchange, Replication stress, BrdU, EdU

1 Introduction

Sister chromatid exchange (SCE) is the exchange event of genetic material between two identical sister chromatid. Sister chromatid exchange was first reported by Taylor et al. with incorporation of tritiated thymidine (^{3}H-TdR) to plant cells [1]. ^{3}H-TdR-incorporated strand presented silver grain dots with autoradiograph images. Although the original experiments were performed to show the Watson-Click semiconservative DNA replication model, clear crossover of sister chromatid was observed simultaneously.

Later, instead, the use of radioisotope differential staining methods was developed with BrdU, which is a thymidine analog [2, 3]. For a long time, SCE methods rely on BrdU-based differential staining method with fluorescence plus Giemsa (FPG) staining or antibody detection methods [4]. Recently, as an alternative of BrdU, EdU-based SCE detection methods have become more popular based on their easy staining protocol [5]. EdU staining is faster than BrdU staining, and chromosome morphology is kept intact as no heat or denature treatment. However, EdU is more

Eisuke Gotoh (ed.), *Chromosome Analysis: Methods and Protocols*, Methods in Molecular Biology, vol. 2519, https://doi.org/10.1007/978-1-0716-2433-3_8,

toxic to cells compared to BrdU. Application of EdU method needs to be optimized by individual researchers in individual laboratory [6].

It has been a long-term discussion how SCE was important for cell. SCE frequency varies in many factors including endogenous and exogenous conditions. Exposure to almost any DNA damaging agents, including ionizing radiation, UV light, alkylating agent, oxidative stress, topoisomerase inhibitors, and crosslinking agents, increases SCE. Therefore, elevation of SCE frequency is a good marker of genotoxic drug screening. SCE analysis with chemicals is often used with S9 fraction to activate pro-mutagenic chemicals. Interestingly, BrdU or EdU itself acts as an inducer of SCE. Many reports showed the clear dose-dependent BrdU induced SCE formation [7, 6]. Even in the minimum low concentration of BrdU, cells still exhibit SCE [8]. Different halogenated pyrimidine induces different levels of SCE [9, 10]. EdU causes SCE more efficiently than BrdU [6]. Therefore, SCE is considered to spontaneously occur.

SCE ratio also varies in different cell lines. Mutations in DNA repair enzymes alter SCE frequency [11]. Loss of homologous recombination activity decreases SCE formation in baseline and/or genotoxic agent-induced SCE. This suggested that homologous recombination repair plays a major role in SCE formation. Other many mutations also increase SCE formation. Nonhomologous end joining (NHEJ) and PARP-deficient cells, Fanconi anemia patient cells increases SCE formation and/or baseline. Also drugs inhibiting these repair pathways increase SCE ratio [12]. However, among these change, two things dramatically changes SCE frequencies. Especially, Bloom syndrome patients, who has mutation in BLM gene, showed almost 10 time increase of SCE [13]. A CHO EM-9, mutant of XRCC1, also showed high induction of SCE as BLM mutations [14].

SCE are observed after adding BrdU for 2 rounds of cell cycles (Fig. 1). In the first S-phase, BrdU is unifilary incorporated in both new synthesized strands. At the first metaphase, both chromatids equally contain BrdU. Therefore, SCE event cannot be visible under microscope at this moment. In the second S-phase, if BrdU is in present, one chromatid contains no-BrdU/BrdU and another chromatid contains bifilary BrdU. Therefore, the second metaphase for the amount of BrdU incorporation is different in paired chromatids. This difference can be differentially stained by FPG. If no BrdU was added in the second strand, one chromatid contains no-BrdU/no-BrdU, and the other contains BrdU/no-BrdU. This is also the differential amount of BrdU in each chromatid as presence or non-presence of BrdU. Although this method needs extra medium change depriving BrdU, clear SCE is visible for anti-BrdU antibody. EdU can be substituted to BrdU with Click reaction in the same manner.

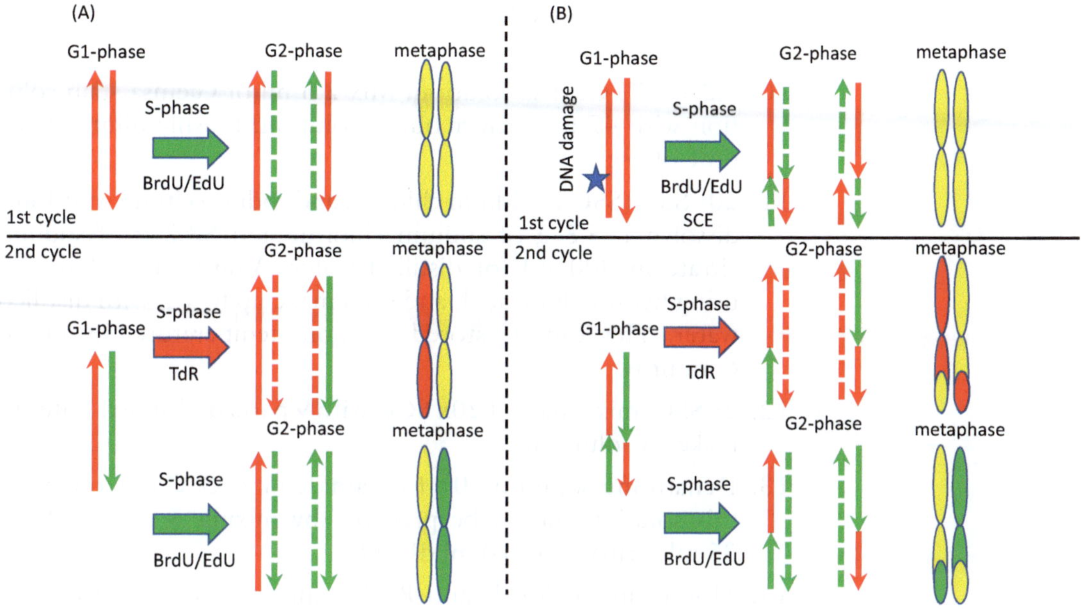

Fig. 1 Diagrammatic illustration of SCE visualization and formation. Red color indicates non-thymidine analog DNA. Green color indicates thymidine analog contained DNA. Yellow color indicates unifilar thymidine analog incorporation. Dashed line indicates newly synthesized DNA strand. (**a**) Differential amounts of thymidine analogs after 2 cycles. (**b**) Visualization and formation of SCE after 2 cycles

2 Materials and Equipment

2.1 Materials

1. Appropriate cell culture media for individual cell such as MEM, alphaMEM, DMEM, and RPMI. Supplemented with fetal bovine serum (FBS) and antibiotic. Prior lot testing is required for FBS. Some lots increase background frequency of SCE formation. Heat inactivation of complement at 56 °C for 20 min is required for FBS. Antibiotics such as Antibiotic-Antimycotic (100x), penicillin-streptomycin solution (5000 U/ml) should be added.
2. 2.1.2 Phosphate buffered saline (−) (PBS).
3. BrdU (5-bromo-2′-deoxyuridine), prepare 10 mM stock solution in 100% DMSO.
4. EdU (5-ethynyl-2′-deoxyuridine), prepare 10 mM stock solution in 100% DMSO.
5. Colcemid solution in PBS (10 μg/mL).
6. Trypsin-EDTA (0.05% or 0.25% Trypsin/EDTA-4Na in PBS).
7. 75 mM hypotonic potassium chloride (KCl).
8. Carnoy's fixative solution (methanol and acetic acid = 3 vol: 1 vol), prepare at time of use.

9. Hoechst 33258 solution. Prepare stock solution in 75 μg/ml in distilled water.
10. 2.2.10 5% Giemsa solution. Mix 2.5 ml of Giemsa stain solution with 47.5 ml Gurr buffer solution. Freshly dilute at the time of use.
11. 20x SSC (SSC = sodium chloride and sodium citrate solution): dissolve 175.3 g of sodium chloride and 88.2 g of sodium citrate in 900 mL of distilled water. Adjust the pH to 7.0 using hydrochloric acid, and then make up to 1 L with distilled water. This can be stored at room temperature for up to 6 months.
12. 2x SSC: mix 1 part of 20x SCC with 9 parts of distilled water to make 2x solution.
13. Denaturing solution 70% formamide in 2x SSC in Coplin jar: this solution should be prepared by mixing 5 ml 20x SSC, 35 ml formamide, 10 ml dH_2O.
14. Mouse monoclonal anti-BrdU antibody (clone BU1/75 (ICR1), BU-1 or B44).
15. 2 N HCl.
16. Alexa488-conjugated goat anti-mouse IgG antibody.
17. EdU staining solution: PBS, 100 mM $CuSO_4$, 1 M sodium ascorbate, Azide Alexa Fluor 488. Mix 200 μL of PBS, 2 μL of $CuSO_4$, 2 μL of sodium ascorbate, 1 μL of Azide Fluor just before use.
18. DAPI in SlowFade.

2.2 Equipment

1. Cell culture facility including biosafety cabinet or clean bench, 37 °C CO_2 controlled incubator, and water bath.
2. Centrifuge, which can rotate 15 ml tubes. No need of temperature control or brake.
3. Aspirator to remove supernatant and trap waste.
4. UV light source such as the transilluminator light box with black light bulbs.
5. Upright light microscope for FPG staining. Fluorescent microscope with appropriate filter sets equipped with CCD camera and software for fluorescent staining with BrdU antibody or EdU Click reaction staining.
6. Slide glass, coverslip, mounting medium.

3 Methods

3.1 Cell Culture

1. G0/G1 synchronized cells can be plated on 6 cm dishes (*see* **Note 1**). If desired, UV, or ionizing radiation treatment, can be applied after cell attachment (*see* **Note 2**).
2. Add BrdU as final concentration of 10 μM. BrdU should be added within a few hours after synchronization is released because some rapidly growing cells may enter S-phase at that time.
3. Let them grow cells for 2 cell cycle times. For CHO with 12 h doubling time, wait for 20 h. For AG1522 with 24 h doubling time, wait 40 h. This timing can be extended based on cell lines and toxicity of testing agents.
4. Colcemid should be added and wait for another 6 h (*see* **Note 3**).
5. Collect media with cells and PBS wash and trypsinized cells into 15 ml tube.
6. Centrifuge cell suspension with 200 g for 5 min at room temperature. Discard supernatant by aspiration or decantation.
7. Add 3 ml of 75 mM KCl hypotonic solution and incubate at 37 °C water bath for 20 min.
8. Prepare Carnoy's fixative solution during this incubation time.
9. Add 1 ml of fixative solution and invert tube and mix well gently.
10. Centrifuge cell suspension with 200 g for 5 min. After centrifugation, larger amount of white pellet should be visible.
11. Discard supernatant by aspiration or decantation. Cell pellet should be loosened by tapping tube.
12. Add 1 mL of fixative, invert tube and mix well, centrifuge, and discard supernatant. Cell pellet should be broken by tapping tube.
13. Repeat this process 2 more times.

    Optional (based on the amount of cells in tubes, added extra fixative solution may be up to 500 μL to adjust cell density on slides).
14. Use yellow pipette tip (100–200 μL), mix well, and break pellet gently with pipetting.
15. Take 40 μL of cell suspended solution and drop onto slides (*see* **Note 4**).
16. Air-dry for 5 min. Do not disturb slides till dry.
17. Slides can be stored for staining after confirmation of metaphase spread quality. Good metaphase spreads do not have

overlapped chromosomes, easily distinguished as a single cell origin metaphase, and no debris of cell membrane or cytoplasm is not visible.

3.2 Staining

3.2.1 Fluorescence Plus Giemsa (FPG) Staining

1. Add 20 μL of Hoechst 33258 dye in PBS solution to slides and put coverslip for 30 min.
2. Place slides on the transilluminator with UVA-UVB lamp and expose 15 min.
3. Put slides into preheated 2xSSC solution at 80 °C for 20 min.
4. Air-dry slides.
5. Put slides into freshly prepared Giemsa solution for 5 min.
6. Rinse slides with running water and air-dry.
7. After mounting media with coverslip, observe under microscope. BrdU/BrdU chromatids should be lighter purple color than BrdU/no-BrdU chromatids (Fig. 2) (*see* **Note 5**).

3.3 BrdU or EdU Uptake for Fluorescent Images

1. It is recommended that BrdU or EdU labeling should be only carried out in first cell cycle. The second cell cycle should be carried out without BrdU/EdU for better staining. With this way, one chromatid contains BrdU/EdU, and another contains no BrdU/EdU. It will give better resolution than 2 cycles labeling with BrdU/EdU, which result in one chromatid containing BrdU/EdU and another containing twice amounts of BrdU/EdU (*see* **Note 6**).
2. For this, asynchronous cell population can be exposed to testing agents, and BrdU/EdU can be added.

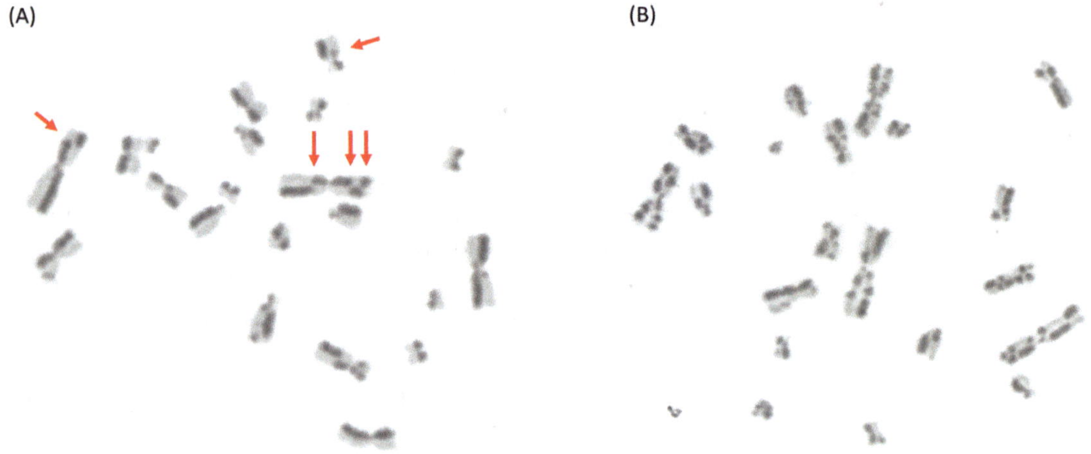

Fig. 2 Fluorescence plus Giemsa (FPG) stained SCE images of CHO wild type (**a**) and EM9 mutants (**b**) without any genotoxic agent treatment. EM9 showed spontaneously elevated SCE frequency. Arrow indicates examples of SCEs

3. Wait for 10 h for rapid growing cell and 20 h for slow growing cells, and then mitotic shake-off can be carried out twice. The first shake should be discarded and fresh media without BrdU/EdU will be added to flasks. One hour later, mitotic shake-off will be carried out, and synchronized G1 cells can be transferred to new dishes. All G1 cells should be labeled with BrdU/EdU once with this method.
4. New dish should be cultured in complete media without labeling.
5. Add colcemid at the appropriate timing and prepare chromosome spreads will be prepared as previously described.

3.4 BrdU Antibody Fluorescent Immunostaining

1. For BrdU antibody staining, slides need to be pretreated with HCl or denatured by heat. 2 N HCl at 37 °C for 3 min treatment along with PBS neutralization is sufficient. 70% formamide in 2x SSC at 80 °C for 2 min and wash with 2x SCC is also sufficient.
2. After blocking with 10% goat serum in PBS, primary antibody mouse BrdU antibody can be applied and followed by washing and secondary anti-mouse IgG antibody conjugated with fluorochrome. 20–50 μL of anti-BrdU antibody (1:500) dilution in 10% goat serum solution should be applied to slides. Parafilm should be placed on slides.
3. Slides will be kept on 37 °C for 1 h. After this, 3 times of washes in PBS for each 10 min.
4. Secondary antibody (1:500) dilution in 10% goat serum solution is applied as the primary antibody. After 1 h incubation, slides will be washed for four times in PBS.
5. 20 μL of mounting media will be applied with coverslip. Observe with fluorescence microscope with appropriate filter. It should be blue and green (Fig. 3).

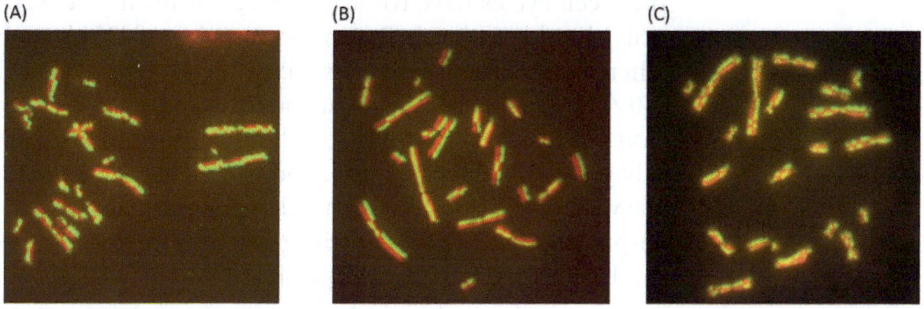

Fig. 3 BrdU antibody or EdU Click chemistry stained SCE images. (**a**) CHO EM9 cells with BrdU staining without any drug treatment to induce SCE. (**b**) CHO PARP-deficient PADR9 with 10 μM EdU treatment and (**c**) 50 μM of EdU treatment. DNA was stained with PI (red) and BrdU/EdU was stained with Alexa488 (green)

3.5 Click Reaction Staining for EdU

1. EdU staining does not need any pretreatment. EdU will be stained with Click reaction between azide and alkyne. Azide is conjugated with Fluor (Alexa488). Ethynyl residue of EdU contains alkyne ($-C \equiv CH$).
2. Prepare EdU staining solution mixture containing 24.4 μL of PBS, 0.25 μL of 100 mM CuSO4, 0.25 μL of 1 M sodium ascorbate, and 0.1 μL of Azide Alexa Fluor per one slide. Apply this solution to slide and put on parafilm slip.
3. Incubated at room temperature for 30 min in dark.
4. Wash with PBS for 10 min.
5. Mount slides with coverslip and DAPI in appropriate anti-fade solution.
6. Observe with fluorescence microscope with appropriate filter. It should be blue and green (Fig. 3).

3.6 Analysis of SCE

1. Find metaphase spreads with SCE under microscope with oil immersion 40x to 100x objective lens. SCE should be scored with selected metaphase spread with nicely spread and minimum chromosome overlapping. SCE should be scored per metaphase spread and analyzed per cell or per chromosome to normalize different chromosome numbers (*see* **Note** 7).
2. For the analysis of FPS staining images, use of camera is not necessarily required. SCE can be scored under microscope. For fluorescent images, each color image needs to be captured by fluorescence microscope equipped with digital camera with appropriate filter sets to process for color composition. Images stored in PC can be analyzed later.

4 Note

1. To obtain clear differentially stained chromosomes, two complete cell cycles have to be occurred for testing cells, or testing cells should undergo 2 complete cycles of DNA replications. Therefore, cells should be synchronized in specific cell cycles. G0/G1 synchronization can be achieved with contact inhibition for normal fibroblast cell lines, mitotic shake-off for rapidly growing attaching cells, and isoleucine depletion or serum starvation methods for general cell lines [15–17]. If cells are suspension culture, centrifugal elutriation may be good method to isolate G1 cells [18].
2. BrdU sensitizes cells to UV and ionizing radiation [17]. Therefore, BrdU should be added after radiation exposure.
3. If mitotic cells are very low frequency, wait for additional few hours.

4. Slides should be cleaned by researchers. Slides can be immersed in iced water. This helps spread metaphase well.

   You can make a single drop at specific location or multiple drops near frosted labeling side and down droplet to the other side.

5. When BrdU concentration is too high, the FPG staining is apparently weak and difficult to be visible. For this, reduce UV exposure time after Hoechst staining. If differential staining is not achieved, it could be problems of staining or not enough time to obtain second cell cycle. Especially after toxic agent treatment, cell cycle time may be extended. In lethal dose treatment, cells do not reach the second mitosis to observe differential staining; therefore, agent treatment should be carried out in sublethal dose range.

6. Concentration of BrdU and EdU is very important. If it is low (<1 μM), the signal is difficult to observe even under fluorescence microscope. If it is too high (>50 μM), BrdU and EdU contribute for significant increase of SCE frequencies (Fig. 3c).

7. SCE frequency in normal mammalian cells should be an average of between 5 and 10 per cell (Fig. 2a). Genotoxic agent exposure increases SCE frequency. If CHO EM9 or cells from Bloom syndrome patients are available, they are good positive control with elevated SCE frequency. Germicidal UV-C lamp in the biosafety cabinet can be a good positive control to induce SCE. It depends on the lamp wattage but has a fluence of a few W/m^2, and 10 J/m^2 exposure (5 second if 2 W/m^2) can give significant increase of SCE. UV-C exposure has to be done for cells attaching in dish. After removing media by aspiration and PBS wash, cells should be exposed without lid. UV-C light from germicidal lamp will not penetrate plastic.

References

1. Taylor JH, Woods PS, Hughes WL (1957) The organization and duplication of chromosomes as revealed by autoradiographic studies using tritium-labeled Thymidinee. Proc Natl Acad Sci U S A 43(1):122–128
2. Perry P, Wolff S (1974) New Giemsa method for the differential staining of sister chromatids. Nature 251(5471):156–158
3. Latt SA (1973) Microfluorometric detection of deoxyribonucleic acid replication in human metaphase chromosomes. Proc Natl Acad Sci U S A 70(12):3395–3399
4. Raza A, Kempski M, Preisler HD, Block AW (1985) A permanent method of detecting Sce by immunofluorescence using monoclonal anti-Brdu antibodies. Cancer Genet Cytogen 15(1–2):187–189
5. Sunada S, Haskins JS, Kato TA (2019) Sister chromatid exchange as a genotoxic stress marker. Methods Mol Biol 1984:61–68. https://doi.org/10.1007/978-1-4939-9432-8_7
6. Haskins JS, Su C, Maeda J, Walsh KD, Haskins AH, Allum AJ, Froning CE, Kato TA (2020) Evaluating the genotoxic and cytotoxic effects of thymidine analogs, 5-Ethynyl-2 -‘-Deoxyuridine and 5-Bromo-2 ‘-Deoxyurdine to mammalian cells. Inter J Molecul Sci 21(18):ARTN 6631. https://doi.org/10.3390/ijms21186631

7. Lambert B, Hansson K, Lindsten J, Sten M, Werelius B (1976) Bromodeoxyuridine-induced sister chromatid exchanges in human lymphocytes. Hereditas 83(2):163–174
8. Pinkel D, Thompson LH, Gray JW, Vanderlaan M (1985) Measurement of sister chromatid exchanges at very low bromodeoxyuridine substitution levels using a monoclonal antibody in Chinese hamster ovary cells. Cancer Res 45(11 Pt 2):5795–5798
9. Heartlein MW, O'Neill JP, Preston RJ (1983) SCE induction is proportional to substitution in DNA for thymidine by CldU and BrdU. Mutat Res 107(1):103–109
10. DuFrain RJ, McFee AF, Linkous S, Jennings CJ, Lowe KW (1984) In vivo SCE analysis using bromodeoxyuridine, iododeoxyuridine, and chlorodeoxyuridine. Mutat Res 139(2): 57–60
11. Wilson DM III, Thompson LH (2007) Molecular mechanisms of sister-chromatid exchange. MutatRes 616(1–2):11–23
12. Hori T (1981) High-incidence of sister chromatid exchanges and chromatid interchanges in the conditions of lowered activity of poly (Adp-ribose) polymerase. Biochem Bioph Res Co 102(1):38–45. https://doi.org/10.1016/0006-291x(81)91485-6
13. Chaganti RS, Schonberg S, German J (1974) Manifold increase in sister chromatid exchanges in blooms syndrome lymphocytes. Proceedings of the National Academy of Sciences of the United States of America 71(11):4508–4512. https://doi.org/10.1073/pnas.71.11.4508
14. Thompson LH, Brookman KW, Dillehay LE, Carrano AV, Mazrimas JA, Mooney CL, Minkler JL (1982) A CHO-cell strain having hypersensitivity to mutagens, a defect in DNA strand-break repair, and an extraordinary baseline frequency of sister-chromatid exchange. MutatRes 95(2–3):427–440
15. Tobey RA, Ley KD (1970) Regulation of initiation of DNA synthesis in Chinese hamster cells. I. Production of stable, reversible G1-arrested populations in suspension culture. JCell Biol 46(1):151–157
16. Nagasawa H, Lin YF, Kato TA, Brogan JR, Shih HY, Kurimasa A, Bedford JS, Chen BP, Little JB (2017) Coordination of the Ser2056 and Thr2609 Clusters of DNA-PKcs in Regulating Gamma Rays and Extremely Low Fluencies of Alpha-Particle Irradiation to G0/G1 Phase Cells. Radiat Res 187(2):259–267. https://doi.org/10.1667/RR14679.1
17. Fujii Y, Genet MD, Roybal EJ, Kubota N, Okayasu R, Miyagawa K, Fujimori A, Kato TA (2013) Comparison of the bromodeoxyuridine-mediated sensitization effects between low-LET and high-LET ionizing radiation on DNA double-strand breaks. Oncol Rep 29(6):2133–2139. https://doi.org/10.3892/or.2013.2354
18. Hinz JM, Yamada NA, Salazar EP, Tebbs RS, Thompson LH (2005) Influence of double-strand-break repair pathways on radiosensitivity throughout the cell cycle in CHO cells. DNA Repair (Amst) 4(7):782–792

Chapter 9

Cytokinesis Blocked Micronuclei Aberration Analysis

Takamitsu A. Kato

Abstract

Cytokinesis blocked micronuclei (CBMN) assay is a rapid and sensitive analysis of chromosome aberrations and miss assortments during cell division. Genotoxic agent exposure produces DNA damage and chromosome fragments. Fragmented chromosomes without centromere failed to attach kinetochore which segregates a pair of homologous chromosomes to each daughter cells at cytokinesis, hence leading to form micronuclei. Chromosome or fragments of chromosome can also form micronuclei when they are not accurately sorted to daughter cells. Using cytochalasin B, an actin inhibitor, blocks cytokinesis of which completion leads serration appearance formed with two daughter cells while nuclei segregation is undergoing. As a result, one cell having two daughter nuclei, i.e., binucleated cell, is produced. By analyzing these binucleated cells, chromosome aberrations can be estimated as well as popular chromosome aberration analysis. Frequency of micronuclei formation predicts the testing agents' genotoxicity. By combining use with centromere-specific probes or DNA damage signal probes, the nature of genotoxicity of tested agents can be estimated.

Key words Cytokinesis blocked micronucleus (CBMN) assay, Cytochalasin B, Nuclear segregation, Genotoxicity

1 Introduction

A micronucleus test is one of cytogenetics analysis to observe chromosome damage. This test is a very robust tool to quantify genomic lesion-induced chromosomal damages produced by testing agents. Compared to metaphase chromosomal aberration analysis, micronuclei are easily identifiable and can analyze much many cells in a short time. Therefore, micronucleus analysis is preferably used for mutagen screening analysis prior to time-consuming labor intense metaphase chromosome analysis [1]. Recently, micronuclei can be digitally recognized with automated microscope and software for much faster analysis.

Eisuke Gotoh (ed.), *Chromosome Analysis: Methods and Protocols*, Methods in Molecular Biology, vol. 2519,
https://doi.org/10.1007/978-1-0716-2433-3_9,

Micronuclei were originally described by Evans et al. in radiation-induced *Vicia faba* roots as a result of chromosome damage [2]. The micronuclei test was developed to observe genotoxic effect in cells after genotoxic agent exposure by Heddle and his collogues [3, 4]. Besides simple Giemsa staining, acridine orange staining was developed by Hayashi and colleagues [5]. However, these original micronuclei analysis has disadvantage against the traditional cytogenetics methods conducted in metaphase chromosome. These metaphase chromosomes analysis clearly indicated damage formed at the initial cell cycle and observed at the time of metaphases; micronuclei were formed at mitosis but analyzed at random time after micronuclei formation. Therefore, it was difficult to link micronucleus phenomenon against chromosome aberrations directly. Addressed to this problem, micronuclei analysis by blocking cytokinesis just after micronuclei formation was reported by Fenech and Morley [6, 7]. Cytokinesis block micronucleus (CBMN) is induced by cytochalasin B, an inhibitor of actin [8]. The advantage of this CBMN method over the traditional micronucleus analysis is distinguishing that whether the cells with micronuclei reached in mitosis or the cells without micronucleus reached the second turn mitosis. Using this methods, accurate micronuclei formation is possible.

Micronuclei are resulted from chromosome breakage and chromosome missegregations (Fig. 1). Orphan chromosome or a part of chromosomes that form micronuclei localizes discrete spot of interphase nucleus. Acentric chromosome fragments cannot be properly transferred to nuclei of daughter cells due to no microtubule attachment resulting in micronuclei. Centric fragments or whole chromosomes can form micronuclei due to failure of proper cytokinesis. Simple Giemsa staining cannot distinguish these differences. However, using antisera from Scleroderma patients containing human autoantibodies, called CREST serum, detects kinetochore structure in micronuclei [9]; hence, FISH (fluorescence in situ hybridization) for centromere [10] and DNA damage can differentiate the origins of micronuclei. For example, use with FISH probes for centromere or CREST serum or kinetochore-specific antibody can identify micronuclei with centromere. Micronuclei with DNA damage signal such as gamma-H2AX can identify broken DNA in micronuclei. These staining techniques allow us to investigate more in depth of mechanisms of micronuclei formation rather than a frequency of micronuclei after chemical exposure (Fig. 1).

This chapter introduces the methods of cytokinesis block micronuclei formation and staining with centromere and DNA damage signals.

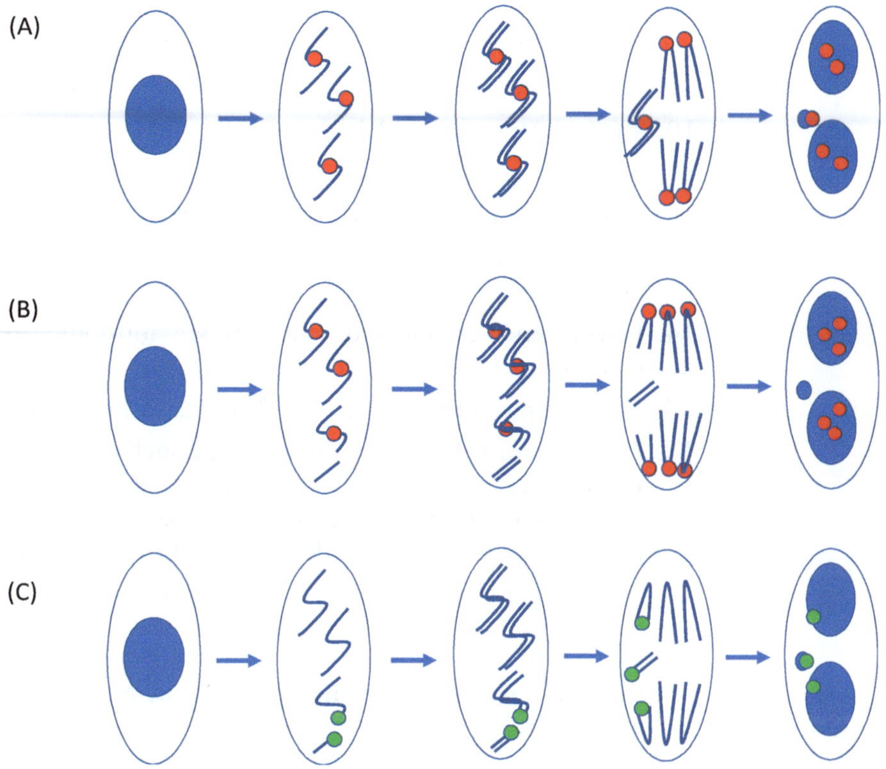

Fig. 1 Processes of micronuclei formation. (**a**) A formation of micronucleus with centromere signal. (**b**) A formation of micronuclei without centromere signal. (**c**) A formation of micronucleus with DNA damage foci. Centromere signals are shown as red dots. DNA damage foci are shown in green dots

2 Materials and Equipment

2.1 Materials

1. Cells: Micronuclei analysis can be done in many cell types and species. Even suspension cells can be used for this technique.
2. Cell culture media: Appropriate cell culture media supplemented with FBS and antibiotics such as heat inactivated 10% FBS and 1% penicillin and streptomycin.
3. PBS (−): Phosphate buffered saline, Mg^{2+}- and Ca^{2+}-free.
4. Trypsin-EDTA: Final concentration of 0.05% (w/v) trypsin, 0.02% (w/v) EDTA (0.53 mM) in Mg^{2+}- and Ca^{2+}-free PBS.
5. 10 mg/ml cytochalasin B: Dissolved in DMSO. Store in −20 °C.
6. 75 mM KCl solution: 2.8 g of KCl should be dissolved in 500 ml of distilled water to make final concertation of 75 mM. Keep this in room temperature.
7. Carnoy's fixative solution: Mix methanol and acetic acid in 3:1 ratio, prepare just before use.

8. 5% Giemsa stain: KaryoMax Giemsa solution should be prepared as 5% solution in Gurr buffer; prepare just before use.
9. Glass slide mounting media.
10. 4% (w/v) paraformaldehyde in PBS.
11. 0.2% TritonX-100 in PBS.
12. Blocking buffer: 10% goat serum in PBS.
13. Anti-gammaH2AX antibody: JBW301 clone from Millipore Sigma, mouse monoclonal antibody.
14. Fluorescence-conjugated anti-mouse antibody: Alexa488-conjugated anti-mouse goat antibody.
15. DAPI in mounting media: 1 μg/ml DAPI in SlowFade.
16. Centromere FISH probe: PNA probe such as AACTAGACA-GAAGCAT or ATTCGTTGGAAACGGGA (alpha satellite centromere) and CENP-B box (ATTCGTTGGAAACGGGA) [11] dissolved in formamide for 50 μM stock solution.
17. 70% (v/v) formamide solution in 2×SSC: Mix 35 ml of formamide, 5 ml of 20×SSC, and 10 ml of water. 20×SSC should be prepared by 175.3 g of NaCl, 88.2 g of Na-Citrate in 900 ml of water. Adjust the pH to 7.0 by NaOH. Make up to 1 l by water.

2.2 Equipment

1. Cell culture facility including biosafety cabinet or clean bench, 37C CO_2 controlled incubator, and water bath.
2. Centrifuge and aspirator.
3. Coplin jar for washing slides.
4. Upright light microscope for Giemsa staining analysis. Fluorescent microscope with appropriate filter sets equipped with CCD camera and software for fluorescent staining.

3 Methods

3.1 The Standard Micronuclei Formation and Giemsa Staining

1. Expose testing agents to cells. Prior to experiments, cells should be cultured on appropriate vessels that can hold enough number of cells such as T25 flask, P35 dish, or 6-well dishes. (*see* **Note 1**).
2. Add 4 μg/ml of cytochalasin B and incubate for appropriate time duration. (*see* **Note 2**).
3. Remove media, wash with PBS, and add trypsin-EDTA to detach cells.
4. Collect detached cells to tube and centrifuge with 300 g for 5 min.
5. Remove supernatant, break cell pellet by tapping, and add 3 mL of 75 mL of KCl.

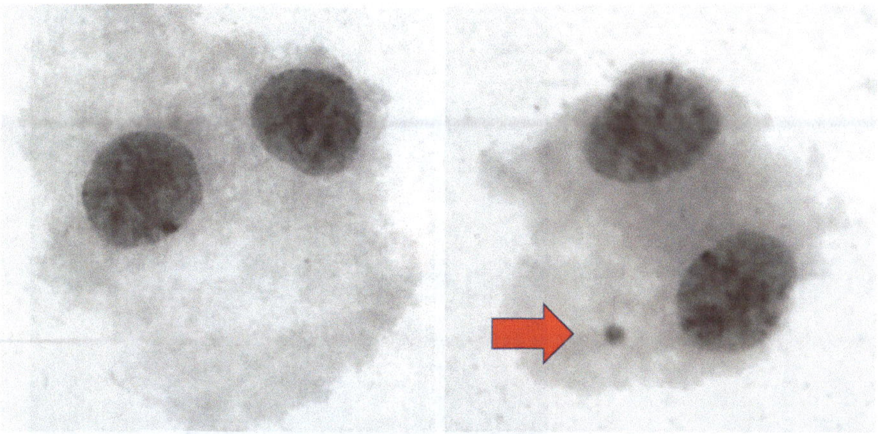

Fig. 2 Binucleated cells with or without micronuclei. Red arrow indicate micronucleus. Dark background shows cytoplasm and discriminates single cells with 2 nuclei or two cells

6. Centrifuge tubes with 300 g for 5 min.
7. Remove supernatant, break cell pellet by tapping, and add 1 mL of Carnoy's fixative solution.
8. Centrifuge tubes with 300 g for 5 min.
9. Remove supernatant, break cell pellet by tapping, add 1 mL of Carnoy's fixative solution.
10. Centrifuge tubes with 300 g for 5 min.
11. Remove supernatant, break cell pellet by pipetting, drop cell solution to the slide glass, and air-dry (*see* **Note 3**).
12. Stain slides with Giemsa solution for 5 min and rinse with tap water and air-dry.
13. Seal slides with coverslip and sealant.
14. Observe under microscope. At least 60×, preferably, 100× objective lens should be used with oil. Only observe binucleated cells for micronuclei analysis (Fig. 2). At least 100 binucleated cells should be analyzed. Preferably, 300 binucleated cells (*see* **Note 4**).

3.2 The Micronuclei Staining with Centromere Staining

1. After the standard micronuclei preparation (Protocol 1), treat slides with 0.001% (w/v) pepsin 50 mM HCl.
2. Add centromere probe solution to slide, put cover slip on this. Centromere probe solution is either pan-centromeric FISH DNA or PNA probe solution or CENP-B Box staining PNA probe. CENP-B box staining probe will not stain Y chromosome.
3. Place slides in humidified chamber like slide box with wet paper towel and incubate for hybridization.
4. After hybridization, wash with PBS for 5 min.

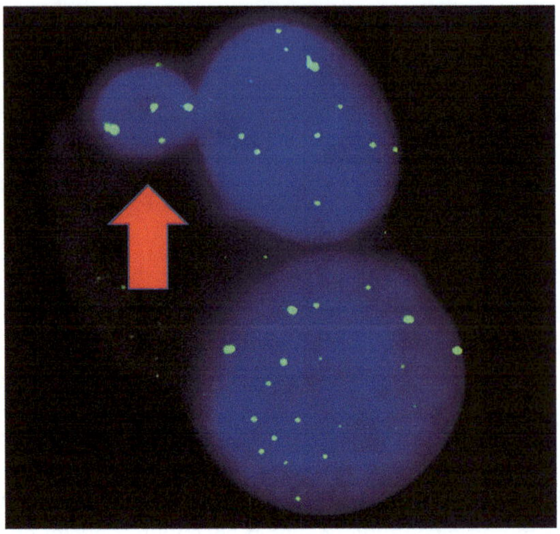

Fig. 3 Binucleated cells with micronucleus and centromere signal. A single cell with 2 nuclei and 1 micronucleus. Red arrow indicates micronucleus and centromere signal (green). Nuclei were counter stained by DAPI (blue). Green dots are centromere signal stained by CREST serum and immunocytochemistry

5. Add DNA counter stain and put coverslip. DAPI (blue), SYBR Green (green), or PI (red) solution should be appropriate. Choose based on filter sets of fluorescent microscope and not overlapping with centromere signals.
6. Observe under fluorescent microscope, and images should be obtained and processed for multicolor images. At least 60×, preferably 100×, objective lens should be used with oil. Individual color images should be obtained first, and then merge them to construct multicolor images (Fig. 3). Obtain frequency of micronuclei with and without centromere signals.

3.3 Micronuclei Staining with Immunocytochemistry

1. Cells should be culture in chamber slide or pre-cleaned coverslip in P35 dish. Coverslip can be sterilized by ethanol, flame, or UV exposure.
2. Cells need to be fixed in 4% paraformaldehyde for 15 min (*see* **Note 5**).
3. Wash cells with PBS and permeabilize cells, and add 0.2% Triton X-100 in PBS solution for 10 min.
4. Remove Triton X-100 by aspirator, and add 50 μL of 10% goat serum in PBS for blocking.
5. Incubate 30 min at room temperature or 4C for overnight.
6. Remove excess goat serum solution and add 30 μL of anti-gammaH2AX antibody solution and cover with parafilm. Place samples on incubator for 1 hour.
7. Wash with PBS for three times.

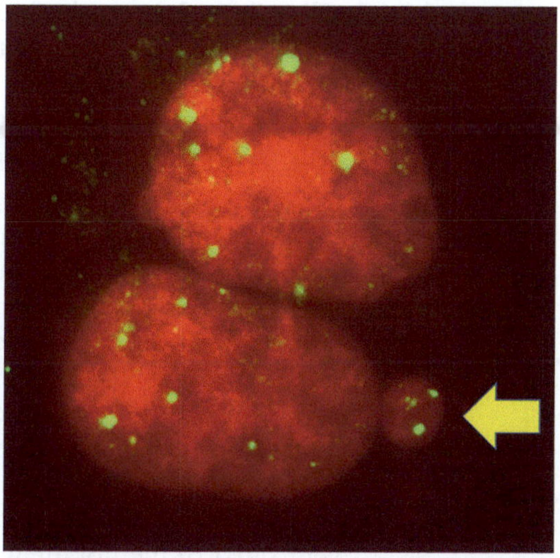

Fig. 4 Micronucleus with gamma-H2AX signal. Binucleated cells show gamma-H2AX signals (green). Nuclei were counter stained by PI (red). A yellow arrow indicates micronuclei with gamma-H2AX foci

8. Add 30 μL of Alexa488-conjugated anti-mouse antibody solution and cover with parafilm.
9. Place samples on incubator for 1 hour.
10. Wash with PBS for three times.
11. Add a drop of DAPI solution and cover with coverslip. If pre-cleaned coverslip is used for cell culture, coverslip should be flipped over slide glass with DAPI.
12. Observe under fluorescent microscope, and images should be obtained and processed for multicolor images. At least 60×, preferably 100×, objective lens should be used with oil. Individual color images should be obtained first, and then merge them to construct multicolor images. Obtain frequency of micronuclei with and without gammaH2AX signals (Fig. 4). gammaH2AX signals may be observed as multiple foci in micronuclei.

4 Notes

1. Cell cycle should be synchronized into specific phases. G0/G1 phases by contact inhibition, serum starvation, or mitotic shake-off methods.
2. Cytochalasin B concentrations should be optimized in the culture conditions. It should be between 2 and 10 μg/ml. Enough to block cytokinesis is required. Exposure genotoxic agents tend to slow down cell cycle. Twenty hours for CHO

cells with 12 hour doubling time, 40 hours for human cells with 24 hours doubling time. If cells are kept too long in cytochalasin B containing media, cells may go to the second mitosis. This will cause four nuclei rather than binucleated cells. Do not analyze cells with four nuclei for micronuclei analysis.

3. Observe the quality of samples under invert phase contrast microscope. The good quality of samples should not have nuclei overlap in cells. Some cytoplasm background will help a discrimination of binucleated cells or two cells. The micronuclei spread preparation is critical for accurate analysis. KCl hypotonic solution treatment prior to fixation should help the quality of micronuclei spread preparation. If cytoplasm is lost after dropping, re-prepare the micronuclei samples by a few drops of formaldehyde to cell sample in fixative solution to prevent burst of membrane when dropped onto slide glass.
4. DNA damaging agents, microtubule inhibitors, and cells with deficiency in DNA repair should observe micronuclei formation. Background micronuclei frequency may vary in cells, but it should be about 0.1–0.3 micronuclei per 100 binucleated cells.
5. Immunocytochemistry needs special fixation method rather than Carnoy's fixative solution.

References

1. Engen A, Maeda J, Wozniak DE, Brents CA, Bell JJ, Uesaka M, Aizawa Y, Kato TA (2015) Induction of cytotoxic and genotoxic responses by natural and novel quercetin glycosides. Mutat Res Genet Toxicol Environ Mutagen 784–785:15–22. https://doi.org/10.1016/j.mrgentox.2015.04.007
2. Evans HJ, Neary GJ, Williamson FS (1959) The relative biological efficiency of single doses of fast neutrons and gamma-rays on Vicia faba roots and the effect of oxygen. Part II. Chromosone damage: the production of micronuclei. Int J Radiat Biol Relat Stud Phys Chem Med 1:216–229. https://doi.org/10.1080/09553005914550311
3. Heddle JA (1973) A rapid in vivo test for chromosomal damage. Mutat Res 18(2):187–190. https://doi.org/10.1016/0027-5107(73)90035-3
4. Countryman PI, Heddle JA (1976) The production of micronuclei from chromosome aberrations in irradiated cultures of human lymphocytes. Mutat Res 41(2–3):321–332. https://doi.org/10.1016/0027-5107(76)90105-6
5. Hayashi M, Sofuni T, Ishidate M Jr (1983) An application of Acridine Orange fluorescent staining to the micronucleus test. Mutat Res 120(4):241–247. https://doi.org/10.1016/0165-7992(83)90096-9
6. Fenech M, Morley AA (1985) Measurement of micronuclei in lymphocytes. Mutat Res 147(1–2):29–36
7. Fenech M, Morley AA (1986) Cytokinesis-block micronucleus method in human lymphocytes: effect of in vivo ageing and low dose X-irradiation. Mutat Res 161(2):193–198. https://doi.org/10.1016/0027-5107(86)90010-2
8. Carter SB (1967) Effects of cytochalasins on mammalian cells. Nature 213(5073):261–264. https://doi.org/10.1038/213261a0
9. Brenner S, Pepper D, Berns MW, Tan E, Brinkley BR (1981) Kinetochore structure, duplication, and distribution in mammalian cells: analysis by human autoantibodies from scleroderma patients. J Cell Biol 91(1):95–102. https://doi.org/10.1083/jcb.91.1.95

10. Becker P, Scherthan H, Zankl H (1990) Use of a centromere-specific DNA probe (p82H) in nonisotopic in situ hybridization for classification of micronuclei. Genes Chromosomes Cancer 2(1):59–62. https://doi.org/10.1002/gcc.2870020111

11. Cartwright IM, Genet MD, Kato TA (2012) A simple and rapid fluorescence in situ hybridization microwave protocol for reliable dicentric chromosome analysis. J Radiat Res. https://doi.org/10.1093/jrr/rrs090

Chapter 10

DNA Damage Foci on Metaphase Chromosomes

Takamitsu A. Kato

Abstract

After DNAs are damaged, DNA repair proteins accumulate and are activated at the DNA damaged site. These accumulated proteins are visualized as foci by fluorescent immunocytochemistry technique. This allows the DNA damage responses in interphase nuclei to be detected; it was earlier times difficult to analyze DNA damage in situ. In order to analyze DNA damage in interphase cells, either DNA is extracted to assay breaks biochemically, or premature chromosome condensation is conducted to observe as chromatin breaks. Although DNA damage-induced foci are typically analyzed in interphase cells, these foci can be also visualized on mitotic chromosomes. The foci where the repair proteins accumulate at the damage site is observed as mitotic chromosome break site. Since mitotic cells attach loosely or not attached to cell culture vessels, it is difficult to analyze foci on chromosomes in culture vessels under a microscope, so metaphase chromosome spread must be prepared for accurate analysis. The cytocentrifuge system is an ideal method to adhere mitotic cells to microscope slides for the fluorescent immunocytochemistry. This chapter introduces cytocentrifuge method to prepare metaphase spread for DNA damage foci analysis.

Key words Cytocentrifugation, Fluorescent immunocytochemistry, DNA damage foci

1 Introduction

The analysis of DNA damage-induced foci is mainly conducted in interphase cells. Many studies are conducted in exponentially growing cell population, but other researches conduct in synchronized cell population or with cell cycle phase-specific markers to investigate cell cycle-dependent effects and to eliminate artifacts incentive from mixed cell cycle population [1, 2]. Since DNA damage foci may correlate with a single DNA damage site [3], the number of DNA damage-induced foci can quantify the number of DNA damage sites with much higher sensitivity than other traditional DNA damage analysis methods. For DNA double-strand break analysis, as low as 1 mGy ionizing radiation dose-induced DNA damage can be detected [4]. This is much lower than the constant or pulse field gel electrophoresis assays which required for at least 5–10 Gy, and

Eisuke Gotoh (ed.), *Chromosome Analysis: Methods and Protocols*, Methods in Molecular Biology, vol. 2519,
https://doi.org/10.1007/978-1-0716-2433-3_10,

neutral comet assay needs a few Gy to detect tail moment, or G1 phase premature chromosome condensation produces approximately 6 PCC fragments per Gy [5].

The commence of DNA damage-induced foci analysis study was gamma-H2AX foci analysis. After gamma-H2AX, many proteins associated with DNA damage responses were reported to form foci. These are including 53BP1, Mre11, Rad50, phospho-NBS, Rad51, pATM, BRCA1, ATR, FANCD2, and others [6–12]. The lack of upstream signals or loss of accessory proteins reduce the formation of foci. For example, cells have paralogs mutations of Rad51, or Rad51 family genes such as Rad51C, Rad51D, XRCC2, XRCC3, or BRCA2 can do form Rad51 foci [13]. Mutations in Fanconi anemia protein such as FANCA, FANCB, FANCC, FANCE, FANCF, and FANCG also cause no formation of FANCD2 foci [14]. Mutations in ATR inhibit MRE11, NBS1, and Rad50 foci formation [15]. MDC1 controls the formation of 53BP1, BRCA1, and MRN foci [16]. 53BP1 is required for the formation of BRCA1 foci [17]. Knockout of H2AX impaired recruitment of NBS1, 53BP1, and BRCA1 foci formation [18]. These foci formation study clarified molecular pathways of signal transduction and cooperation of DNA damage-responding proteins.

Although many studies have been conducted on the interphase cells, foci analysis on metaphase chromosomes is still limited [19, 20]. The standard metaphase spread preparation protocol uses the traditional Carnoy's fixative of mixture of 3:1 methanol and acetic acid following hypotonic solution treatment with 0.075 M KCl. Carnoy's fixative solution definitely affects epitopes of antibody recognition sites or foci formation by strong acidity. As a result, foci is not usually detectable after regular Carnoy's fixative solution. Therefore, alternative fixation methods are required to stain immunocytochemistry. The standard interphase cell foci analysis methods use paraformaldehyde and permeabilize by detergent or methanol/acetone fixation. Although this fixation protocol is good to preserve cellular and chromosome structure, the metaphase chromosome analysis is needed nevertheless, and cellular membrane must be removed for good chromosome analysis. Therefore, in order to achieve foci staining on metaphase chromosomes, spreading metaphase followed by immunocytochemical fixation without use of Carnoy's fixative step must be required. The cytocentrifugation method, which is a system to adhere cells to microscope slides, can prepare metaphase spread without prefixation. Once metaphase spread is prepared, the regular fixation and staining process can be performed. This chapter introduces the use of cytocentrifuge followed by fixation for immunocytochemistry. This allows the researchers to be able to visualize DNA damage-induced foci on metaphases chromosomes.

2 Materials and Equipment

2.1 Materials

1. Cells: DNA damage foci analysis can be done in many cell types and species. Even suspension cells can be used for this technique.
2. Cell culture media: Appropriate cell culture media such as MEM, MEMalpha, DMEM, RPMI, or F-12 supplemented with FBS and antibiotics such as heat-inactivated 10% FBS and 1% penicillin and streptomycin.
3. PBS (phosphate-buffered saline, Mg^{2+}- and Ca^{2+}-free).
4. Trypsin-EDTA, final concentration of 0.05% (w/v) Trypsin, 0.02% (w/v) EDTA (0.53 mM) in Mg^{2+}- and Ca^{2+}-free PBS.
5. Colcemid: 10 μg/ml stock solution in PBS or Hank's balanced salt solution. Store in 4 °C.
6. Hypotonic solution: KCL, 2.8 g of KCl should be dissolved in 500 ml of distilled water to make final concertation of 75 mM. Keep this in room temperature.
7. 4% paraformaldehyde solution in PBS.
8. Permeabilization solution: KCM solution (120 mM KCl, 20 mM NaCl, 0.5 mM EDTA, 10 mM Tris HCl pH 7.5 with 0.1% Triton X-100).
9. Blocking agent: 10% goat serum in PBS should be prepared in sterile manner.
10. Anti-gH2AX antibody: JBW301 clone from Millipore Sigma, mouse monoclonal antibody.
11. Fluorescence conjugated anti-mouse antibody: Alexa488-conjugated anti-mouse goat antibody.
12. DAPI: 1 μg/ml DAPI in SlowFade solution, store in 4 °C.

2.2 Equipment

1. Cell culture facility including biosafety cabinet or clean bench, 37 °C CO_2 controlled incubator, and water bath.
2. Centrifuge, which can rotate 15 ml tubes. No need of temperature control or brake.
3. Aspirator to remove supernatant and trap waste.
4. Cytocentrifuge such as cytospin.
5. Coplin jar.
6. Slide glass, coverslip.
7. Fluorescent microscope with appropriate filter sets equipped with CCD camera and software.

3 Methods

3.1 DNA Damage-Induced Foci Metaphase by Cytocentrifugation and Fixation

1. Cells plated in appropriate vessels such as dish, flask, or tube are treated with testing agents.
2. After certain incubation time followed by colcemid treatment, single-cell suspension should be prepared (*see* **Note 1**).
3. Cells should be centrifuged, and cytocentrifugation system should be prepared by putting slide glass into a centrifugation holder.
4. Appropriate amount of cell solution should be added and centrifuged with 800 rpm for 3 min (*see* **Note 2**).
5. Remove slides from cytospin and rinse with PBS in Coplin jar, and then put them into 4% paraformaldehyde in Coplin jar for 15 min at room temperature (*see* **Note 3**).
6. Wash slides with PBS and put them into the permeabilization solution for 10 min.
7. Put 50 μL of blocking solution and cover it with parafilm. Incubate them in the moist chamber overnight in a fridge.

3.2 Staining and Analysis

1. Prepare primary antibody solution by diluting 1:300 in 10% goat serum solution. Add antibody solution to slide and cover with parafilm. Incubate for 1 hour at 37 °C incubator. 20–50 μl of antibody solution per slide should be sufficient.
2. Wash with PBS in Coplin jar for 5 min three times.
3. Prepare secondary antibody solution by diluting 1:500. Add antibody solution to slide and cover with parafilm. Incubate for 1 hour at 37 °C in the incubator.
4. Wash with PBS in Coplin jar for 5 min three times.
5. Apply one drop of DAPI solution to slide and seal it with coverslip.
6. Observe under fluorescent microscope with appropriate filters.
7. Obtain individual color images and merge them with software.
8. Analyze foci number and distribution.
9. Figure 1 shows the typical images; from left to right image, numbers of damaged foci (green signal) increased as DNA damages increased.

4 Notes

1. For rapidly growing cells, 3 hours should be sufficient for colcemid treatment.

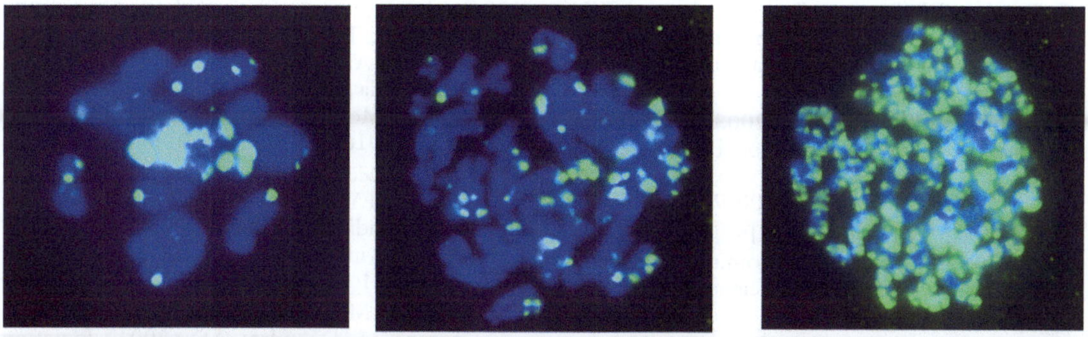

Fig. 1 Representative images of DNA damage foci on metaphase chromosome. Green signal indicates gamma-H2AX and blue signal indicates DAPI

For slowly growing cells, 6 hours should be sufficient. Culturing cells with colcemid for extended time will cause high background and cellular damages.

2. Cytology funnels with filter card system or CytoSep MEGA funnel system can be used. Appropriate amounts of cells should be loaded into the cytocentrifuge. Cytology funnels may take a few hundred μL of sample solution. MEGA funnel system may take a few ml of sample solution with more area. Centrifugation speed and time may be varied in the systems and cell lines.

   This method needs good metaphase spread by cytocentrifugation. Therefore, optimizations of speed and time of centrifugation are necessary. Even in the best conditions, some metaphase chromosomes may overlap, good metaphase spreads have to be searched on the slides.

   Too much speed will cause burst cell and chromosomes. It will cause severe morphological aberrations.

3. After this step, rinse with ice-cold methanol may help cell adhesion of slides.

4. The number of foci should be increased with DNA damage. Those foci tend to be larger and greater signal than foci on interphase nuclei. If DNA damage occurs in interphase, the number of foci on metaphase may be affected by DNA repair and checkpoint [20].

References

1. Kato TA, Nagasawa H, Weil MM, Little JB, Bedford JS (2006) Levels of gamma-H2AX Foci after low-dose-rate irradiation reveal a DNA DSB rejoining defect in cells from human ATM heterozygotes in two at families and in another apparently normal individual. RadiatRes 166(3):443–453
2. Genet SC, Fujii Y, Maeda J, Kaneko M, Genet MD, Miyagawa K, Kato TA (2012) Hyperthermia inhibits homologous recombination repair and sensitizes cells to ionizing radiation in a time and temperature dependent manner. J Cell Physiol. https://doi.org/10.1002/jcp.24302

3. Rogakou EP, Boon C, Redon C, Bonner WM (1999) Megabase chromatin domains involved in DNA double-strand breaks in vivo. J Cell Biol 146(5):905–915
4. Rothkamm K, Lobrich M (2003) Evidence for a lack of DNA double-strand break repair in human cells exposed to very low x-ray doses. Proc Natl Acad Sci U S A 100(9):5057–5062
5. Cornforth MN, Bedford JS (1983) X-ray-induced breakage and rejoining of human interphase chromosomes. Science 222(4628): 1141–1143
6. Schultz LB, Chehab NH, Malikzay A, Halazonetis TD (2000) p53 binding protein 1 (53BP1) is an early participant in the cellular response to DNA double-strand breaks. J Cell Biol 151(7):1381–1390. https://doi.org/10.1083/jcb.151.7.1381
7. Maser RS, Monsen KJ, Nelms BE, Petrini JH (1997) hMre11 and hRad50 nuclear foci are induced during the normal cellular response to DNA double-strand breaks. Mol Cell Biol 17(10):6087–6096. https://doi.org/10.1128/mcb.17.10.6087
8. Haaf T, Golub EI, Reddy G, Radding CM, Ward DC (1995) Nuclear foci of mammalian Rad51 recombination protein in somatic cells after DNA damage and its localization in synaptonemal complexes. Proc Natl Acad Sci U S A 92(6):2298–2302. https://doi.org/10.1073/pnas.92.6.2298
9. Gatei M, Young D, Cerosaletti KM, Desai-Mehta A, Spring K, Kozlov S, Lavin MF, Gatti RA, Concannon P, Khanna K (2000) ATM-dependent phosphorylation of nibrin in response to radiation exposure. Nat Genet 25(1):115–119. https://doi.org/10.1038/75508
10. Davalos AR, Kaminker P, Hansen RK, Campisi J (2004) ATR and ATM-dependent movement of BLM helicase during replication stress ensures optimal ATM activation and 53BP1 focus formation. Cell Cycle 3(12): 1579–1586. https://doi.org/10.4161/cc.3.12.1286
11. Gatei M, Zhou BB, Hobson K, Scott S, Young D, Khanna KK (2001) Ataxia telangiectasia mutated (ATM) kinase and ATM and Rad3 related kinase mediate phosphorylation of Brca1 at distinct and overlapping sites. In vivo assessment using phospho-specific antibodies. J Biol Chem 276(20):17276–17280. https://doi.org/10.1074/jbc.M011681200
12. Garcia-Higuera I, Taniguchi T, Ganesan S, Meyn MS, Timmers C, Hejna J, Grompe M, D'Andrea AD (2001) Interaction of the Fanconi anemia proteins and BRCA1 in a common pathway. Mol Cell 7(2):249–262. https://doi.org/10.1016/s1097-2765(01)00173-3
13. Yuan SS, Lee SY, Chen G, Song M, Tomlinson GE, Lee EY (1999) BRCA2 is required for ionizing radiation-induced assembly of Rad51 complex in vivo. Cancer Res 59(15): 3547–3551
14. Siddique MA, Nakanishi K, Taniguchi T, Grompe M, D'Andrea AD (2001) Function of the Fanconi anemia pathway in Fanconi anemia complementation group F and D1 cells. Exp Hematol 29(12):1448–1455. https://doi.org/10.1016/s0301-472x(01)00754-8
15. Franchitto A, Pichierri P (2002) Bloom's syndrome protein is required for correct relocalization of RAD50/MRE11/NBS1 complex after replication fork arrest. J Cell Biol 157(1):19–30. https://doi.org/10.1083/jcb.200110009
16. Stewart GS, Wang B, Bignell CR, Taylor AM, Elledge SJ (2003) MDC1 is a mediator of the mammalian DNA damage checkpoint. Nature 421(6926):961–966. https://doi.org/10.1038/nature01446
17. Wang B, Matsuoka S, Carpenter PB, Elledge SJ (2002) 53BP1, a mediator of the DNA damage checkpoint. Science 298(5597):1435–1438. https://doi.org/10.1126/science.1076182
18. Celeste A, Petersen S, Romanienko PJ, Fernandez-Capetillo O, Chen HT, Sedelnikova OA, Reina-San-Martin B, Coppola V, Meffre E, Difilippantonio MJ, Redon C, Pilch DR, Olaru A, Eckhaus M, Camerini-Otero RD, Tessarollo L, Livak F, Manova K, Bonner WM, Nussenzweig MC, Nussenzweig A (2002) Genomic instability in mice lacking histone H2AX. Science 296(5569):922–927
19. Chan KL, Palmai-Pallag T, Ying S, Hickson ID (2009) Replication stress induces sister-chromatid bridging at fragile site loci in mitosis. Nat Cell Biol 11(6):753–760. https://doi.org/10.1038/ncb1882
20. Kato TA, Okayasu R, Bedford JS (2009) Signatures of DNA double strand breaks produced in irradiated G1 and G2 cells persist into mitosis. J Cell Physiol 219(3):760–765. https://doi.org/10.1002/jcp.21726

Chapter 11

FISH with Whole Chromosome Painting Probes

Takamitsu A. Kato

Abstract

Some types of chromosome aberrations are not easily identified by the traditional Giemsa staining. It usually needs some banding technique and skilled person's eye. Whole chromosome painting FISH probe can stain designated entire chromosomes or domains in metaphase chromosomes or interphase nuclei, respectively. It allows to visually identify translocations, deletions, or amplifications of specific chromosomes. Once chromosomes are stained, even non-skilled researchers can easily identify those chromosome aberrations. Whole chromosome painting FISH has higher sensitivity to detect chromosome aberrations. This chapter introduces the method for whole chromosome painting FISH staining.

Key words FISH, Whole chromosome painting, Chromosome domain, Chromosome aberration

1 Introduction

FISH (fluorescence in situ hybridization) staining with the whole chromosome painting probes makes possible to identify the specific chromosomes by coloring the target chromosomes in the metaphase spreads. This painted chromosomes are visually distinguished, and it is much easier than traditional Giemsa stain by which morphological differences such as chromosome length or length ratio of short arm and long arm. The identified chromosomes are one pair in the normal somatic cells, but when chromosome aberrations occur by genotoxic stress or carcinogenesis steps, FISH signals may increase more than 2 as fragmented or amplified chromosomes or decrease by loss of chromosomes. The analysis of chromosome aberrations needs special training and experience. It is hard to identify some types of chromosome aberrations not only by the traditional Giemsa staining but even also by technical demanding banding method. The sensitivity of fluorescence signals is much higher than banding pattern. Therefore, FISH with whole chromosome painting probes has advantages over the traditional banding pattern in particular to detect small insertions and deletions.

Eisuke Gotoh (ed.), *Chromosome Analysis: Methods and Protocols*, Methods in Molecular Biology, vol. 2519,
https://doi.org/10.1007/978-1-0716-2433-3_11,

The basic technique of FISH is in situ hybridization (ISH) originally reported by Pardue and Gall in 1969 [1]. Complement DNA sequences were hybridized to target sequences in an in situ manner. Early ISH protocols used radioisotopes to detect signals [2, 3]. Radiolabeled probes were replaced to biotylated probes [4], and then fluorescence probes were introduced in 1980 [5]. FISH technique rapidly became popular, and two-color FISH and multicolor FISH were continually developed soon after [6, 7]. FISH technique was expanded widely. DNA analogous probes such as PNA (peptide nucleic acid) were used for their higher affinity [8]. Combination of multicolor system enabled to distinguish individual chromosome pairs in different color as multicolor FISH (mFISH) [7]. mBAND technique, a chromosome with multicolor banding staining, allows inversion and deletion detection [9]. Site directional probe for stain one of chromatids allows to detect inversion aberration and telomere aberrations [10]. In diagnosis, amplification and deletion of specific gene locus such as HER2 were done with FISH analysis [11].

This chapter introduces FISH technique with whole chromosome painting probes. This is relatively easy and cost-effective to detect chromosome aberrations. Whole chromosome painting analysis can detect chromosome breaks and translocation. Chromosome breaks simply increase the number of detectable chromosomes with signals more than 2. Translocation can be identified as fused chromosomes with mixture of stained and non-stained area. Translocation is usually difficult to be detected by Giemsa stain, especially complex-type translocation, which is generated by more than three chromosomes. These complex-type translocations often occurred after ionizing radiation exposure and carcinogenesis step. Whole chromosome painting FISH can even easily distinguish these types of aberrations.

2 Materials and Equipment

2.1 Materials

1. Cells: Whole chromosome painting FISH analysis can be done in many cell types and species as long as specific probes are available.
2. Cell culture media: Appropriate cell culture media supplemented with FBS and antibiotics such as heat-inactivated 10% FBS and 1% penicillin and streptomycin.
3. PBS (−): phosphate-buffered saline, Mg^{2+}- and Ca^{2+}-free.
4. 0.05% (w/v) trypsin, 0.02% (w/v) EDTA (0.53 mM) in PBS (−).
5. 10 μg/ml Karyomax Colcemid solution.
6. 75 mM KCl solution.

7. Carnoy fixative solution: Mix methanol and acetic acid in 3:1 ratio; just prepare before use.
8. 70% formamide solution in 2xSSC: Mix 35 ml of formamide, 5 ml of 20xSSC, and 10 ml of water.
9. 50% formamide, 2xSCC, 20% dextran sulfate for hybridization solution.
10. Whole chromosome painting probes, for example, Cambio star*FISH biotin or FITC labeled.
11. Alexa 594 conjugated streptavidin for biotin-labeled probes.
12. Alexa 488 conjugated anti-fluorescein goat antibody for FITC-labeled probes.
13. 1 μg/ml DAPI in SlowFade.

2.2 Equipment

1. Cell culture facility including biosafety cabinet or clean bench, 37 °C CO_2 controlled incubator, and water bath.
2. Centrifuge and aspirator.
3. Coplin jar for washing slides.
4. Fluorescent microscope with camera, software.

3 Methods

3.1 The Standard Metaphase Chromosome Spread Preparation

1. Prior to experiments, cells should be cultured on appropriate vessels that can hold enough number of cells such as T25 flask, P35 dish, or 6-well dishes. Prudentially, cell cycle should be synchronized into specific phases. G0/G1 phases by contact inhibition, serum starvation, or mitotic shake-off methods.
2. Expose testing agents to cells and release them from G0/G1 phase.
3. Add 0.1 μg/ml of Colcemid at the appropriate timing and incubate for appropriate time duration. Approximately 2–6 h of Colcemid treatment should be enough.
4. Transfer media to tube, add PBS and transfer to tube, and add trypsin-EDTA to detach cells. Collect detached cells to tube and centrifuge with 300 g for 5 min.
5. Remove supernatant, break cell pellet by tapping, and add 3 mL of 75 mM KCl.
6. Incubate tube in the water bath at 37 °C for 20 min.
7. Add 1 mL of Carnoy fixative solution, and centrifuge with 300 g for 5 min.
8. Remove supernatant, break cell pellet by tapping, and add 1 mL of Carnoy fixative solution. Centrifuge tubes with 300 g for 5 min.

9. Repeat this three times.
10. Remove supernatant, break cell pellet by pipetting, drop cell solution to the slide glass, and air dry. Observe the quality of samples under invert phase contrast microscope (*see* **Note 1**).

3.2 Whole Chromosome Painting

1. Denature a biotin-labeled human chromosome 1 specific painting probe and a FITC-conjugated probe for human chromosome 2 (Cambio, Cambridge, UK) at 84 °C for 14 min and incubate at 37 °C for 45 min (*see* **Note 2**).
2. Immerse slides in the preheated 73 °C denature solution in Coplin jar for 3 min (*see* **Note 3**).
3. Dehydrate slides with ethanol series of 70%, 90%, and 100% ethanol in Coplin jar.
4. Probe solution should be added to slides. Put coverslip and seal it by rubber cement or nail polish.
5. Keep slides in 37 °C overnight.
6. The slides should be washed with PN buffer at 45 °C for 10 min followed by rinsing with PBS.
7. Add 10% goat serum in PBS for blocking.
8. Add solution of Alexa 594 conjugated streptavidin and Alexa 488 conjugated anti-fluorescein antibody to enhance FISH signal. Incubate for 1 h at 37 °C (*see* **Note 4**).
9. Wash slides with PBS for three times.
10. Add DAPI solution and put coverslip.
11. Observe under fluorescent microscope, and images should be obtained and processed for multicolor images. At least 60x, preferably 100x objective lens should be used with oil. Individual color images should be obtained first, and then merge them to construct multicolor images (Fig. 1). Obtain frequency of chromosome aberrations (*see* **Note 5**).

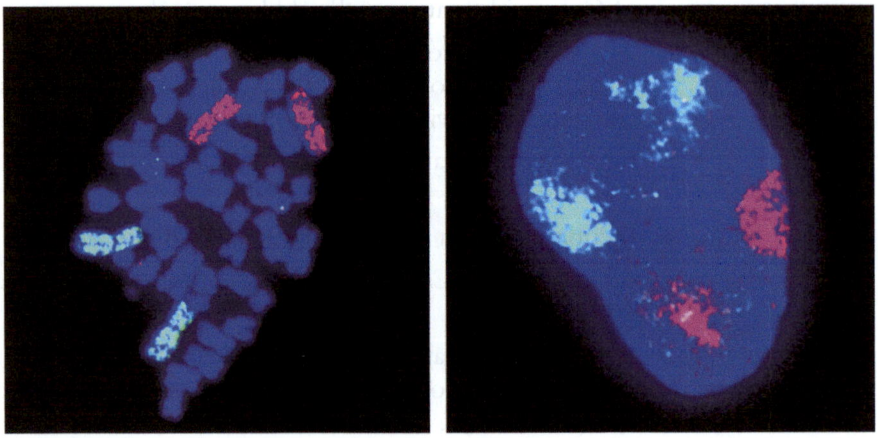

Fig. 1 Chromosome 1 and 2 painting of metaphase chromosomes and interphase cells. Chromosome 1 is red, and chromosome 2 is green. Blue is DAPI nuclear signal

4 Notes

1. The good quality of samples should not have chromosomes overlap or no visible cytoplasm or membrane. Too much destruction of metaphase chromosomes may cause trouble. If there are too many cells on slide, dilute cell solution with Carnoy fixative solution.
2. Temperature control can be operated by PCR machine. Probes may be mixed in hybridization buffer (50% formamide, 2xSCC, 20% dextran sulfate).
3. Heating denature in formamide may destruct chromosome structure. In that case, heating temperature and duration can be shortened, or slides can be pretreated with 4% paraformaldehyde to maintain structure. It is essential to preheat denature solution. Keep Coplin jar at least 30 min and check temperature by thermometer. Do not put too many slides in Coplin jar for denature. Each time a slide is placed, temperature will be dropped.
4. Amplifications by biotin/streptavidin or fluorescein/antibody are recommended to get consistent good signals.
5. Whole chromosome painting probe will stain designated chromosomes. Normal cells should be stained for a pair of chromosomes except sex chromosomes. With DNA damage, chromosomes may be broken and translocated with other chromosomes [12]. Whole chromosome painting of FISH can detect translocations, deletions, and amplifications.

References

1. Pardue ML, Gall JG (1969) Molecular hybridization of radioactive DNA to the DNA of cytological preparations. Proc Natl Acad Sci U S A 64(2):600–604. https://doi.org/10.1073/pnas.64.2.600
2. John HA, Birnstiel ML, Jones KW (1969) RNA-DNA hybrids at the cytological level. Nature 223(5206):582–587. https://doi.org/10.1038/223582a0
3. Buongiorno-Nardelli M, Amaldi F (1970) Autoradiographic detection of molecular hybrids between RNA and DNA in tissue sections. Nature 225(5236):946–948. https://doi.org/10.1038/225946a0
4. Manning JE, Hershey ND, Broker TR, Pellegrini M, Mitchell HK, Davidson N (1975) A new method of in situ hybridization. Chromosoma 53(2):107–117. https://doi.org/10.1007/BF00333039
5. Bauman JG, Wiegant J, Borst P, van Duijn P (1980) A new method for fluorescence microscopical localization of specific DNA sequences by in situ hybridization of fluorochromelabelled RNA. Exp Cell Res 128(2):485–490. https://doi.org/10.1016/0014-4827(80)90087-7
6. Nederlof PM, van der Flier S, Wiegant J, Raap AK, Tanke HJ, Ploem JS, van der Ploeg M (1990) Multiple fluorescence in situ hybridization. Cytometry 11(1):126–131. https://doi.org/10.1002/cyto.990110115
7. Schrock E, du Manoir S, Veldman T, Schoell B, Wienberg J, Ferguson-Smith MA, Ning Y, Ledbetter DH, Bar-Am I, Soenksen D, Garini Y, Ried T (1996) Multicolor spectral karyotyping of human chromosomes. Science 273(5274):494–497. https://doi.org/10.1126/science.273.5274.494

8. Hultdin M, Gronlund E, Norrback K, Eriksson-Lindstrom E, Just T, Roos G (1998) Telomere analysis by fluorescence in situ hybridization and flow cytometry. Nucleic Acids Res 26(16):3651–3656. https://doi.org/10.1093/nar/26.16.3651
9. Hada M, Gersey B, Saganti PB, Wilkins R, Cucinotta FA, Wu H (2010) mBAND analysis of chromosome aberrations in human epithelial cells induced by gamma-rays and secondary neutrons of low dose rate. Mutat Res 701(1): 67–74. https://doi.org/10.1016/j.mrgentox.2010.03.009
10. Robinson E, McKenna MJ, Bedford JS, Goodwin EH, Cornforth MN, Bailey SM, Ray FA (2019) Directional genomic hybridization (dGH) for detection of Intrachromosomal rearrangements. Methods Mol Biol 1984: 107–116. https://doi.org/10.1007/978-1-4939-9432-8_13
11. Kallioniemi OP, Kallioniemi A, Kurisu W, Thor A, Chen LC, Smith HS, Waldman FM, Pinkel D, Gray JW (1992) ERBB2 amplification in breast cancer analyzed by fluorescence in situ hybridization. Proc Natl Acad Sci U S A 89(12):5321–5325
12. Kato TA, Okayasu R, Jeggo PA, Fujimori A (2011) ASPM influences DNA double-strand break repair and represents a potential target for radiotherapy. Int J Radiat Biol 87(12): 1189–1195. https://doi.org/10.3109/09553002.2011.624152

Chapter 12

Telomere Aberration Detection by PNA FISH Probe

Takamitsu A. Kato

Abstract

Telomere is a structure of the end cap of chromosomes. Telomere gets shorter as cell aging and progressing cell division. Shorter telomere may cause telomere fusion, thus inducing genomic instability. Telomere dysfunction can be visualized by PNA FISH probe against telomere repeat sequence (TTAGGG)n. PNA probes have higher hybridization affinity than DNA probes. The traditional FISH or modified FISH protocol can stain telomere relatively easier than whole-chromosome painting probes. This chapter introduces PNA telomere FISH protocol to visualize telomere signals on metaphase chromosomes.

Key words Telomere, Telomeric repeat sequences, End cap of chromosomes, Heyflick's limit, Telomere fusion, PNA FISH

1 Introduction

Telomere, a component of protein and DNA, is a structure located on the end cap of eukaryote chromosomes that protects chromosomes by capping them [1]. In early studies, Herman Muller identified stability of chromosome end region against radiation and named it telomere [2]. Barbara McClintock investigated chromosome breakage fusion theory with telomere [3]. Cells lose a short part of telomere in every cell division, so aged cells or senescence cells have shorter telomere length than youth cell. Telomere length is believed to be important for maintenance of genome. Cells with short telomere are associated with crisis of cell division, which is well known as Heyflick's limit [4]. Telomere length can be elongated by telomerase identified by Greider and Blackburn [5] or alternative lengthening telomere [6, 7]. Only specific types of cells including stem cells or cancer cells maintain certain length of telomere during cell proliferation.

Since telomere ends are structurally similar to DNA double strand breaks, telomere forms specific structure to prevent from recognized as DNA damage [8]. This structure is called T-loop. When telomere is getting too short, telomere no longer can

Eisuke Gotoh (ed.), *Chromosome Analysis: Methods and Protocols*, Methods in Molecular Biology, vol. 2519, https://doi.org/10.1007/978-1-0716-2433-3_12,

maintain T-loop structure. This shortening or loss of telomere may cause DNA fusion events of two telomere ends. Telomere fusion can be inter-chromosome fusion or intra-chromosomal chromatid fusion with or without telomere. These telomere fusion may cause genomic instability [9]. Telomere can be also damaged exogenously by genotoxic agent exposure or endogenously by reactive oxidative species, and defect of DNA repair and telomere breaks and telomere exchange can happen [10, 11]. Some of these damages can be detected with strand-specific telomere FISH method.

These events can be visualized with telomere-specific probe FISH. Telomere is highly conserved repeated sequence of (TTAGGG)n among vertebrates; the same sequence of probe can be used for other species. Telomere FISH to human chromosomes was first reported in 1988 [12]. Later, PNA (peptide nucleic acid) probe was introduced for better efficiency of hybridization [13]. Telomere aberrations are commonly seen in many cancerous cells [14]. This chapter introduces simple telomere staining on metaphase chromosomes.

2 Materials and Equipment

2.1 Materials

1. Cells: Telomere aberration analysis can be done with many cell types with the same probes because telomere sequence is highly conserved in species.
2. Cell culture media: Appropriate cell culture media supplemented with FBS and antibiotics such as heat-inactivated 10% FBS and 1% penicillin and streptomycin.
3. PBS: phosphate-buffered saline, Mg^{2+}- and Ca^{2+}-free.
4. 0.05% (w/v) trypsin and 0.02% (w/v) EDTA (0.53 mM) in Mg^{2+}- and Ca^{2+}-free PBS.
5. 10 μg/ml Karyomax Colcemid solution.
6. 75 mM KCl: 2.8 g of KCl dissolved in 500 ml of distilled water to make final concertation of 75 mM. Keep this in room temperature.
7. Carnoy fixative solution: Mix methanol and acetic acid in 3:1 ratio just before use.
8. 4% paraformaldehyde.
9. 0.2% TritonX-100 in PBS.
10. 0.002% (w/v) pepsin solution in 50 mM HCl: Mix 50 ml of 50 mM HCl and 0.1 ml of 1% (w/v) pepsin. Use freshly prepared solution. 1% pepsin should be stored in −20 °C.
11. 70% formamide solution in 2xSSC: Mix 35 ml of Formamide, 5 ml of 20xSSC, and 10 ml of water.

12. Telomere probe: PNA probe of FAM conjugated $(CCCTAA)_3$ or FAM conjugated $(TTAGGG)_3$, 50 μM stock solution in formamide.
13. Telomere staining solution: 200 nM of telomere probe, 20 mM Tris-HCl, pH 7.4, and 60% formamide. For 20 μL solution, mix 4 μL of 100 mM Tris-HCl, 12 μL of formamide, 0.2 μL of probe, and 4 μL of water.
14. 1 μg/ml DAPI in SlowFade.

2.2 Equipment

1. Cell culture facility including biosafety cabinet or clean bench, 37 °C CO_2 controlled incubator, and water bath.
2. Centrifuge and aspirator.
3. Coplin jar for washing slides.
4. PCR machine to denature probes.
5. Fluorescent microscope with camera and software.

3 Methods

3.1 The Standard Metaphase Chromosome Spread Preparation

1. Prior to experiments, cells should be cultured on appropriate vessels that can hold enough number of cells such as T25 flask, P35 dish, or 6-well dishes.
2. Expose testing agents to cells.
3. Add 0.1 μg/ml of Colcemid and incubate for appropriate time duration. Approximately 2–6 h of Colcemid treatment should be enough.
4. Transfer media to tube, add PBS and transfer to tube, and add trypsin-EDTA to detach cells. Collect detached cells to tube and centrifuge with 300 g for 5 min.
5. Remove supernatant, break cell pellet by tapping, and add 3 mL of 75 mM KCl.
6. Incubate tube in the water bath at 37 °C for 20 min.
7. Add 1 mL of Carnoy fixative solution and centrifuge with 300 g for 5 min.
8. Remove supernatant, break cell pellet by tapping, and add 1 mL of Carnoy fixative solution. Centrifuge tubes with 300 g for 5 min.
9. Repeat this three times.
10. Remove supernatant, break cell pellet by pipetting, drop cell solution to the slide glass, and air dry. Observe the quality of samples under invert phase contrast microscope (*see* **Note 1**).

3.2 Telomere Staining

1. Slides should be placed onto 4% Paraformaldehyde in PBS in Coplin Jar for 15 Min
2. Rinse with PBS.
3. Put slides into 0.002% (w/v) pepsin 50 mM HCl at 37 °C for 15 min (*see* **Note 2**).
4. Rinse with water.
5. Dehydrate with ethanol series of 70%, 90%, and 100% ethanol in Coplin jar.
6. By using PCR machine, denature telomere probe solution at 85 °C for 5 min and then cool to 37 °C till adding solution to slides.
7. Put slides in a 2x SSC 70% formamide solution in Coplin jar at 80 °C for 2 min.
8. Dehydrate with ethanol series of 70%, 90%, and 100% ethanol in Coplin jar.
9. Add telomere probe solution to slide and put cover slip on this.
10. Place slides in humidified chamber like slide box with wet paper towel and incubate for staining at 37 °C overnight.
11. Wash with PBS for 5 min and remove coverslip during washing.
12. Add DNA counter stain (DAPI in SlowFade) and put coverslip.
13. Observe under fluorescent microscope, and images should be obtained and processed for multicolor images. At least 60x, preferably 100x objective lens should be used with oil.

    Individual color images should be obtained first, and then merge them to construct multicolor images (Fig. 1).
14. Obtain frequency of telomere aberration per metaphase chromosomes. Telomere aberrations include loss of telomere signals, interstitial telomere signals, and telomere amplifications (*see* **Note 3**).

4 Notes

1. The good quality of samples should not have chromosomes overlap or no visible cytoplasm or membrane. Too much destruction of metaphase chromosomes may cause trouble. If there are too many cells on slide, dilute cell solution with Carnoy fixative solution. Good telomere staining may need prestaining procedure. It is necessary to have nice metaphase spread without cytoplasm and membrane. We used pepsin in HCl solution to remove any remnants, but other variation such as detergent treatment may help.
2. If RNA is troubled with background signals, treat slides with RNase A for 30 min prior this step.

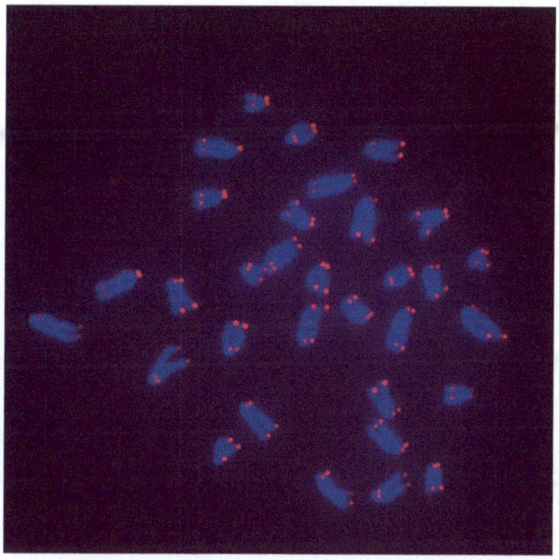

Fig. 1 PNA FISH staining for telomere. Metaphase spread of mouse fibroblast cell was stained by PNA probe (red). DNA stain by DAPI (blue)

3. Telomere should be visible at the end of chromosomes (Fig. 1). But telomere signal may be visible at interstitial region. Interstitial telomere signals may be originated from chromosome fusions or amplifications of telomere region [14]. Strength of telomere signals is associated with telomere size on the chromosomes. Senescence cells or highly divided cells may have shorter telomeres. Mouse has longer telomere than human, so using cells originated from mouse may be beneficial for a positive control. Telomere fusion and break may occur when telomere got damaged.

References

1. Blackburn EH, Gall JG (1978) A tandemly repeated sequence at the termini of the extrachromosomal ribosomal RNA genes in Tetrahymena. J Mol Biol 120(1):33–53. https://doi.org/10.1016/0022-2836(78)90294-2
2. Muller HJ (1938) The remaking of chromosomes. Collecting Net 13:182–198
3. McClintock B (1941) The stability of broken ends of chromosomes in Zea Mays. Genetics 26(2):234–282
4. Hayflick L, Moorhead PS (1961) The serial cultivation of human diploid cell strains. Exp Cell Res 25:585–621. https://doi.org/10.1016/0014-4827(61)90192-6
5. Greider CW, Blackburn EH (1985) Identification of a specific telomere terminal transferase activity in Tetrahymena extracts. Cell 43(2 Pt 1):405–413. https://doi.org/10.1016/0092-8674(85)90170-9
6. Bryan TM, Englezou A, Gupta J, Bacchetti S, Reddel RR (1995) Telomere elongation in immortal human cells without detectable telomerase activity. EMBO J 14(17):4240–4248
7. Bryan TM, Englezou A, Dalla-Pozza L, Dunham MA, Reddel RR (1997) Evidence for an alternative mechanism for maintaining telomere length in human tumors and tumor-derived cell lines. Nat Med 3(11):1271–1274. https://doi.org/10.1038/nm1197-1271
8. van Steensel B, Smogorzewska A, de Lange T (1998) TRF2 protects human telomeres from end-to-end fusions. Cell 92(3):401–413. https://doi.org/10.1016/s0092-8674(00)80932-0

9. Riboni R, Casati A, Nardo T, Zaccaro E, Ferretti L, Nuzzo F, Mondello C (1997) Telomeric fusions in cultured human fibroblasts as a source of genomic instability. Cancer Genet Cytogenet 95(2):130–136. https://doi.org/10.1016/s0165-4608(96)00248-8
10. Bailey SM, Meyne J, Chen DJ, Kurimasa A, Li GC, Lehnert BE, Goodwin EH (1999) DNA double-strand break repair proteins are required to cap the ends of mammalian chromosomes. Proc Natl Acad Sci U S A 96(26): 14899–14904. https://doi.org/10.1073/pnas.96.26.14899
11. Bailey SM, Cornforth MN, Kurimasa A, Chen DJ, Goodwin EH (2001) Strand-specific postreplicative processing of mammalian telomeres. Science 293(5539):2462–2465. https://doi.org/10.1126/science.1062560
12. Moyzis RK, Buckingham JM, Cram LS, Dani M, Deaven LL, Jones MD, Meyne J, Ratliff RL, Wu JR (1988) A highly conserved repetitive DNA sequence, (TTAGGG)n, present at the telomeres of human chromosomes. Proc Natl Acad Sci U S A 85(18):6622–6626
13. Nielsen PE, Egholm M, Berg RH, Buchardt O (1991) Sequence-selective recognition of DNA by strand displacement with a thymine-substituted polyamide. Science 254(5037): 1497–1500. https://doi.org/10.1126/science.1962210
14. Maeda J, Yurkon CR, Fujisawa H, Kaneko M, Genet SC, Roybal EJ, Rota GW, Saffer ER, Rose BJ, Hanneman WH, Thamm DH, Kato TA (2012) Genomic instability and telomere fusion of canine osteosarcoma cells. PLoS One 7(8):e43355. https://doi.org/10.1371/journal.pone.0043355

Chapter 13

Nontraditional Method for Telomere Staining by PNA Probes

Takamitsu A. Kato

Abstract

The standard FISH uses DNA probes to hybridize to the designated complementary strands. This is DNA-DNA interaction, and it usually takes much longer time to obtain detectable signals compared to other reactions such as immunochemical reactions and simple chemical reactions. Certain proteins bind to specific DNA sequences and regulate the biological function of DNA. These DNA-binding proteins have specific domains to interact with single- or double-stranded DNA. Some of telomere proteins apparently bind to telomere sequence and form nucleoprotein complex to protect chromosome ends. Using telomere PNA probes, probes can be accumulated at the telomere sites in a non-hybridization manner. This chapter introduces nontraditional PNA telomere staining protocol without DNA-DNA hybridization to visualize telomere locations on metaphase chromosomes.

Key words Telomere, PNA probes, Non-nucleic acid hybridization method, Nucleoprotein complex

1 Introduction

Telomere, a component of protein and DNA, is a structure located on the end cap of eukaryote chromosomes and protects chromosomes by capping them [1]. Since the end of chromosomes with telomere sequence is similar to DNA double strand breaks, telomere forms shelterin to prevent from recognized as DNA damage [2]. Shelterin coordinates the formation of T-loop. T-loops are maintained by a series of telomere-associated proteins such as TREF1, TERF2, POT1, TIN2, RAP1, and TPP1. TREF1 and TREF2 bind to the double-stranded TTAGGG region of telomere [3, 4]. RAP1 is associated with TREF2 [5]. POT1 binds to single-stranded DNA [6]. TPP1 is associated with POT1 [7]. TIN2 is associated with other shelterin proteins [7]. Any loss of these proteins induces telomere dysfunctions.

FISH is a method of detecting and studying DNA sequences in chromosomes. Telomere is highly conserved repeated sequence of $(TTAGGG)_n$ in vertebrates. Therefore, complementary sequence can be used for hybridization for FISH. Telomere FISH to human

Eisuke Gotoh (ed.), *Chromosome Analysis: Methods and Protocols*, Methods in Molecular Biology, vol. 2519,
https://doi.org/10.1007/978-1-0716-2433-3_13,

chromosomes was first reported in 1988 [8]. Later, PNA (peptide nucleic acid) probe was introduced for better efficiency of hybridization [9]. Telomere aberrations are commonly seen in cancerous cells [10]. Since telomere sequence is highly conserved in vertebrate cells, the same sequence of probe can be used for other species. FISH provides highly specific signal, but it requires harsh denature process and long-term incubation period to obtain signals. However, using specific PNA probes, telomere signal can be detectable without denaturing. Although the mechanism of this staining may be probe and protein reaction rather than probe and DNA reaction, this method can omit several steps necessary in the traditional FISH methods. This method relies on specific DNA sequence decoy to fish DNA binding proteins, and complimentary sequence is not as efficient as another. Telomere probe of $(CCCTAA)_3$ provides much stronger telomere signals than that of $(TTAGGG)_3$ [11]. Several proteins are reported to bind specifically to the single-strand $(CCCTAA)_n$ telomere motif [12]. Since regular FISH probes are not recognized by proteins as targets, they cannot be used for this method. Another example of this method is CENP-B box sequence for centromere staining [13]. CENP-B box sequence can be used not only FISH probe but also direct staining method [11]. Using the technique here, DNA binding proteins and their targets can be revealed to understand their functions [14].

2 Materials and Equipment

2.1 Materials

1. Cells: Telomere aberration analysis can be done with many cell types with the same probes because telomere sequence is highly conserved in species.
2. Cell culture media: Appropriate cell culture media supplemented with FBS and antibiotics such as heat-inactivated 10% FBS and 1% penicillin and streptomycin.
3. PBS: phosphate-buffered saline, Mg^{2+}- and Ca^{2+}-free.
4. 0.05% (w/v) trypsin, 0.02% (w/v) EDTA (0.53 mM) in Mg^{2+}- and Ca^{2+}-free PBS.
5. 10 μg/ml Karyomax Colcemid solution.
6. 75 mM KCl: 2.8 g of KCl dissolved in 500 ml of distilled water to make final concertation of 75 mM. Keep this in room temperature.
7. Carnoy fixative solution: Mix methanol and acetic acid in 3:1 ratio just before use.
8. 4% Paraformaldehyde.
9. 0.2% TritonX-100 in PBS.

10. 0.002% (w/v) pepsin solution in 50 mM HCl: Mix 50 ml of 50 mM HCl and 0.1 ml of 1% (w/v) pepsin. Use freshly prepared solution. 1% pepsin should be stored in −20 °C.
11. 70% formamide solution in 2xSSC: Mix 35 ml of formamide, 5 ml of 20xSSC, and 10 ml of water.
12. Telomere probe: PNA probe of FAM conjugated $(CCCTAA)_3$. 50 μM stock solution in formamide.
13. Telomere staining solution: 200 nM of telomere probe, 20 mM Tris-HCl, pH 7.4, and 60% formamide. For 20 μL solution, mix 4 μL of 100 mM Tris-HCl, 12 μL of formamide, 0.2 μL of probe, and 4 μL of water.
14. 1 μg/ml DAPI in SlowFade.

2.2 Equipment

1. Cell culture facility including biosafety cabinet or clean bench, 37 °C CO_2 controlled incubator, and water bath.
2. Centrifuge and aspirator.
3. Coplin jar for washing slides.
4. Fluorescent microscope with camera, software.

3 Methods

3.1 The Standard Metaphase Chromosome Spread Preparation

1. Prior to experiments, cells should be cultured on appropriate vessels that can hold enough number of cells such as T25 flask, P35 dish, or 6-well dishes.
2. Expose testing agents to cells.
3. Add 0.1 μg/ml of Colcemid and incubate for appropriate time duration. Approximately 2–6 h of Colcemid treatment should be enough.
4. Transfer media to tube, add PBS and transfer to tube, and add trypsin-EDTA to detach cells. Collect detached cells to tube and centrifuge with 300 g for 5 min.
5. Remove supernatant, break cell pellet by tapping, and add 3 mL of 75 mM KCl.
6. Incubate tube in the water bath at 37 °C for 20 min.
7. Add 1 mL of Carnoy fixative solution and centrifuge with 300 g for 5 min.
8. Remove supernatant, break cell pellet by tapping, and add 1 mL of Carnoy fixative solution. Centrifuge tubes with 300 g for 5 min.
9. Repeat this three times.
10. Remove supernatant, break cell pellet by pipetting, drop cell solution to the slide glass, and air dry. Observe the quality of samples under invert phase contrast microscope (*see* **Note 1**).

3.2 Telomere Staining

1. Slides should be placed onto 4% Paraformaldehyde in PBS in Coplin Jar for 15 Min
2. Rinse with PBS.
3. Put slides into 0.002% (w/v) pepsin 50 mM HCl at 37 °C for 15 min (*see* **Note 2**).
4. Rinse with water.
5. Dehydrate with ethanol series of 70%, 90%, and 100% ethanol in Coplin jar. Prepare telomere staining solution during dehydration.
6. Add telomere probe solution to slide and put cover slip on this.
7. Place slides in humidified chamber like slide box with wet paper towel and incubate for staining at 37 °C overnight.
8. Wash with PBS for 5 min and remove coverslip during washing.
9. Add DNA counter stain (DAPI in SlowFade) and put coverslip.
10. Observe under fluorescent microscope, and images should be obtained and processed for multicolor images. At least 60x, preferably 100x objective lens should be used with oil.

    Individual color images should be obtained first, then merge them to construct multicolor images (Fig. 1).
11. Obtain frequency of telomere aberration per metaphase chromosomes. Telomere aberrations include loss of telomere signals, interstitial telomere signals, and telomere amplifications (*see* **Note 3**).

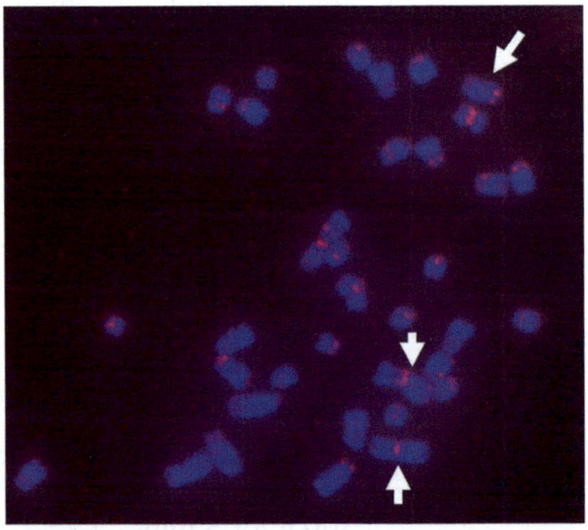

Fig. 1 PNA non-traditional FISH staining for telomere. Metaphase spread of fibroblast cell was stained by PNA probe (red). DNA stain by DAPI (blue). Arrows show telomere-telomere fusion regions which seem like as centromere

4 Notes

1. The good quality of samples should not have chromosomes overlap or no visible cytoplasm or membrane. Too much destruction of metaphase chromosomes may cause trouble. If there are too many cells on slide, dilute cell solution with Carnoy fixative solution. Good telomere staining may need prestaining procedure. It is necessary to have nice metaphase spread without cytoplasm and membrane. We used pepsin in HCl solution to remove any remnants, but other variation such as detergent treatment may help.
2. If RNA is troubled with background signals, treat slides with RNase A for 30 min prior this step.
3. Telomere should be visible at the end of chromosomes (Fig. 1). But telomere signal may be visible at interstitial region (Fig. 1). Interstitial telomere signals may be originated from chromosome fusions or amplifications of telomere region [10]. Strength of telomere signals is associated with telomere size on the chromosomes. Senescence cells or highly divided cells may have shorter telomeres. Telomere fusion and break may occur when telomere got damaged. Mouse and canine cancer cells can be a good positive control to observe telomere fusion.

References

1. Blackburn EH, Gall JG (1978) A tandemly repeated sequence at the termini of the extrachromosomal ribosomal RNA genes in Tetrahymena. J Mol Biol 120(1):33–53. https://doi.org/10.1016/0022-2836(78)90294-2
2. van Steensel B, Smogorzewska A, de Lange T (1998) TRF2 protects human telomeres from end-to-end fusions. Cell 92(3):401–413. https://doi.org/10.1016/s0092-8674(00)80932-0
3. Smith S, de Lange T (1997) TRF1, a mammalian telomeric protein. Trends Genet 13(1):21–26. https://doi.org/10.1016/s0168-9525(96)10052-4
4. Broccoli D, Smogorzewska A, Chong L, de Lange T (1997) Human telomeres contain two distinct Myb-related proteins, TRF1 and TRF2. Nat Genet 17(2):231–235. https://doi.org/10.1038/ng1097-231
5. Li B, Oestreich S, de Lange T (2000) Identification of human Rap1: implications for telomere evolution. Cell 101(5):471–483. https://doi.org/10.1016/s0092-8674(00)80858-2
6. Baumann P, Cech TR (2001) Pot1, the putative telomere end-binding protein in fission yeast and humans. Science 292(5519):1171–1175. https://doi.org/10.1126/science.1060036
7. Jones M, Bisht K, Savage SA, Nandakumar J, Keegan CE, Maillard I (2016) The shelterin complex and hematopoiesis. J Clin Invest 126(5):1621–1629. https://doi.org/10.1172/JCI84547
8. Moyzis RK, Buckingham JM, Cram LS, Dani M, Deaven LL, Jones MD, Meyne J, Ratliff RL, Wu JR (1988) A highly conserved repetitive DNA sequence, (TTAGGG)n, present at the telomeres of human chromosomes. Proc Natl Acad Sci U S A 85(18):6622–6626
9. Nielsen PE, Egholm M, Berg RH, Buchardt O (1991) Sequence-selective recognition of DNA by strand displacement with a thymine-substituted polyamide. Science 254(5037):1497–1500. https://doi.org/10.1126/science.1962210
10. Maeda J, Yurkon CR, Fujisawa H, Kaneko M, Genet SC, Roybal EJ, Rota GW, Saffer ER,

Rose BJ, Hanneman WH, Thamm DH, Kato TA (2012) Genomic instability and telomere fusion of canine osteosarcoma cells. PLoS One 7(8):e43355. https://doi.org/10.1371/journal.pone.0043355

11. Genet MD, Cartwright IM, Kato TA (2013) Direct DNA and PNA probe binding to telomeric regions without classical in situ hybridization. Mol Cytogenet 6(1):42. https://doi.org/10.1186/1755-8166-6-42
12. Marsich E, Piccini A, Xodo LE, Manzini G (1996) Evidence for a HeLa nuclear protein that binds specifically to the single-stranded d (CCCTAA)n telomeric motif. Nucleic Acids Res 24(20):4029–4033. https://doi.org/10.1093/nar/24.20.4029
13. Masumoto H, Masukata H, Muro Y, Nozaki N, Okazaki T (1989) A human centromere antigen (CENP-B) interacts with a short specific sequence in alphoid DNA, a human centromeric satellite. J Cell Biol 109(5): 1963–1973
14. Dickey TH, Altschuler SE, Wuttke DS (2013) Single-stranded DNA-binding proteins: multiple domains for multiple functions. Structure 21(7):1074–1084. https://doi.org/10.1016/j.str.2013.05.013

Chapter 14

Visualizing Active Replication Regions in S-Phase Chromosomes

Eisuke Gotoh

Abstract

A basic question of cell biology is how DNA folds to chromosome. A number of recently accumulated evidences have suggested that folding of chromosome proceeds tightly coupled with DNA replication progresses. Drug-induced PCC is a useful tool for visualization of the interphase nuclei, in particular, S-phase, as S-phase prematurely condensed chromosomes (S-phase PCC). Active replicating DNA is labeled directly with Cy3-dUTP by bead loading method, and then S-phase nuclei is immediately condensed prematurely by calyculin A to obtain S-phase PCC. Active replicating regions on S-PCC are observed under a scanning confocal microscope. Cy3-dUTP-labeled S-phase PCCs clearly reveal the drastic transitional change of chromosome formation through S-phase, starting from a "cloudy nebula" to numerous numbers of "beads on a string" and finally to "striped arrays of banding structured chromosome" known as G- or R-banding pattern. The number, distribution, and shape of replication foci were also measured in individual subphase of S-phase; maximally ~1400 foci of 0.35 μm average radius size were scored at the beginning of S-phase, and the number is reduced to ~100 at the end of S-phase. Drug-induced PCC clearly provided the new insight that eukaryote DNA replication is tightly coupled with the chromosome condensation/compaction for construction of eukaryote higher-ordered chromosome structure.

Key words Chromosome condensation/compaction, Chromosome structure, DNA replication, Replication foci, Premature chromosome condensation (PCC), Prematurely condensed chromosomes (PCCs), Calyculin A, Chromosome territories, Beads loading method, Cy3-dUTP

1 Introduction

Chromosome condensation/compaction is a basic and profound principle of cell biology. Very recently accumulated evidences have implied that eukaryote DNA replication/transcription is involved in formation of chromosomes [1–3], however how it involves remains unclear.

Eukaryote DNA replication commences at thousands of multiple replication origins at the entry of S-phase, and then replication proceeds spatially and temporally synchronizing in an ordered fashion through S-phase. Cytogenetic observation of the replicating

Eisuke Gotoh (ed.), *Chromosome Analysis: Methods and Protocols*, Methods in Molecular Biology, vol. 2519, https://doi.org/10.1007/978-1-0716-2433-3_14,

DNA in interphase nucleus is a most direct approach for elucidating the dynamics of folding DNA to chromosome along with replication progresses. In the very earlier studies, the fiber autoradiography of DNA labeled with [^{3}H]-thymidine revealed the synchronized synthesizing replicon of size ranging from 50 to 300 kbp [4–6]. Then, the thymidine analogue BrdU (bromodeoxyuridine) has been introduced to localize and measure the replication foci [7–9] (*see* **Note 1**). More recently, the replication regions and chromosome formation in living cells were visualized using Cy5-dUTP directly labelled fluorescent DNA [2]. The recent development of a laser scanning microscopy (LSM) and new fluorophore such as Cy3 and Cy5 has much improved the resolution of the observed DNA replication regions (*see* **Notes 2** and **3**). The accumulated findings of dynamic organization of replication factory in the nuclei have raised interest in replication DNA and how it folds within interphase nuclei during S-phase and how the folding process leads to higher-ordered whole chromosome. A number of supporting findings have actually shown the tight joining of DNA replication and chromosome conformation [1–3]. However, these results were derived from the observation in interphase nuclei; thus the resulting resolution and information were limited because chromosomes are only visible in the mitotic stage but are invisible as decondensed structure in interphase [2, 10].

Premature chromosome condensation (PCC) is a useful and unique technique that allows the interphase nuclei to be visualized like as a condensed form of mitotic chromosome [10–12]; thus a lot of studies including DNA replication and DNA repair have been performed using the PCC method [13–20]. Conventional PCC has been carried out by cell fusion using either fusogenic viruses [21] or polyethylene glycol [22] (cell fusion-mediated PCC), but use of these mediator is usually problematic, and resulting chromosomes are mixture of those of inducer and recipient cells [10]. These drawbacks of the technique have been recently overcome with a much easier and more rapid technique using calyculin A or okadaic acid, specific inhibitors of protein phosphatases (drug-induced PCC technique) [10, 23–26]. Drug-induced PCC is now becoming much more popular and has been used in a wide range of cytogenetic applications [23–25, 27–36].

The morphology of prematurely condensed chromosomes (PCCs) of the interphase nucleus varies during the cell cycle as the time of PCC induction either by cell fusion or drugs [10, 21, 23]. G1-, S-, or G2-PCCs show the distinctive shape: (i) G1-phase nuclei produced univalent (i.e., un-replicated) chromosomes (G1-PCCs), (ii) S-phase nuclei condensed to a "pulverized" appearance that a mixture of univalent (i.e., un-replicated) and bivalent (i.e., replicated) chromosomes (S-PCCs), and (iii) G2-phase nuclei showed bivalent chromosomes that are similar to mitotic chromosomes, although less condensed (G2-PCCs)

[10]. This cell cycle-dependent morphological transition is typically seen through S-phase: (i) in early S, thin filamentous fibers spread widely; (ii) in middle S, "pulverized" appearance of chromosomes consists of thin and thick chromatin fiber; and (iii) in late S, most of chromosomes condensed to show bivalent chromosomes like mitotic chromosomes [14, 21]. These different characteristic appearances seen in different subphases of S-PCCs are useful for identifying the involvement of chromosome formation and DNA replication in individual subphase of S-phase; thus PCC is a possible way to answer the above question.

In the present study, we utilized the drug-induced PCC technique. Cy3-dUTP directly labeled fluorescent DNA of individual subphase of S-phase nuclei was condensed prematurely, and the dynamics of chromosome condensation/compaction and the distribution of replication foci on chromosomes were analyzed. The results strongly suggested that DNA replication and chromosome condensation/compaction are tightly coupled in constructing the higher-ordered structure of eukaryote chromosome.

2 Materials

1. Human normal karyotype fibroblast cell line GM05389: Distributed from Coriell Cell Repositories (Camden, NJ) (*see*, **Note 4**).
2. 100 μM calyculin A ($C_{50}H_{81}N_4O_{15}P$): 100 μg of calyculin A, lyophilized in a vial shipped from the supplier, dissolved with 1 mL of solvent (100% DMSO or 100% ethanol, calyculin A does not dissolve in water) will make ~100 μM stock solution (i.e., M.W. of calyculin A is 1009.17). Aliquot in 100 μL in a microcentrifuge tube and store at −20 °C and stable at least 3 months. For the more details, *see* Chapter 5.
3. 10 μM Cy3-dUTP (M.W. 1362.29): 100 nmol concentration in shipping, dilute to final concentration in MEM medium. Cy3-dUTP is pH insensitive.
4. Glass Beads. Sizes of 425–600 μm diameter, acid-washed glass beads (cat. No. G8772) were from Sigma-Aldrich (St. Louise, MO) (*see* **Note 5**).
5. 0.075 M potassium chloride (KCl): for hypotonic cell treatment, dissolve 7.45 g KCl (M.W. 74.55) in 100 mL Milli-Q water or RO (reverse osmotic) water to make a 1 M KCl stock solution. Take 7.5 ml of the 1 M stock solution and make it up to 100 mL with Milli-Q water or RO water to obtain a 0.075 M KCl working solution. Usually it does not need sterilizing.

6. Carnoy's fixative: Take three parts of 100% methanol and one part of 100% acetic acid (e.g., 150 mL of methanol and 50 mL of acetic acid to make 200 mL fixative) and mix together well. After mixing, the temperature will be increased according to exothermic reaction. It is preferable to prepare just before use; however, the fixative can be stored at −20 °C for a month (*see* **Note 6**).
7. 200 nM of DAPI (M.W. 277.324) working solution: Dilute with PBS for counterstaining DNA/chromosomes.

3 Methods

3.1 Cell Culture and Cy3-dUTP Labeling by a Bead Loading Method

1. Cy3-dUTP was loaded in the cells using a bead loading procedure as described previously [2, 37, 38]. The procedure facilitates incorpolate Cy3-dUTP into cell nucleus within very short time following transient permeabilization of the cell membranes [2]. Human fibroblast cells GM05389 were seeded on a 35 mm culture dish with 0.17-mm-thick glass base (Iwaki, Japan) specially designed for high magnification observation under inverted confocal laser-scanning microscopes. Cells were maintained exponentially and asynchronously growing in MEM medium supplemented with 15% fetal calf serum at 37 °C in 5% CO_2 atmosphere with 95% humidification.
2. Culture medium was replaced with 37 °C prewarmed 50 μl of 10 μM Cy3-dUTP in MEM medium, 425–600-μm-diameter glass beads were sprinkled onto the cells, tapping and rocking the dish several times, and then the beads were rinsed off with prewarmed PBS (37 °C). It usually took a couple of minutes for complete the loading procedure (*see* **Note 7**).
3. Prewarmed (37 °C) MEM was added to the culture and incubated for another 10 min after starting of Cy3-dUTP loading (*see* **Notes 8** and **9**).

3.2 Premature Chromosome Condensation

1. Immediately after the Cy3-dUTP loading procedure was completed, chromosomes were condensed prematurely with 50 nM of calyculin A. Premature chromosome condensation (PCC) using calyculin A was carried out as described elsewhere [23, 25, 38, 39] except that the incubation time was kept as short as possible (within 10 min) to obtain a high spatial and temporal resolution; otherwise replication foci can be merged together resulting in a less spatial/temporal resolution (*see* **Notes 1** and **2**). For cytogenetic scoring, 30 min incubation was used to obtain sufficient number of chromosomes [40–42], but chromosomes begin to condense within the first 5 min of incubation [40], and a substantial number of chromosomes condensed after 10 min incubation (*see* **Note 8**).

2. After incubation with calyculin A, cells were harvested, swollen in 0.075 M KCl for 10 min at 37 °C, fixed three times with prechilled Carnoy's fixative, dropped on a glass slide, and air-dried (*see* **Note 6**). Finally, DNA was counterstained with 200 ng/ml of DAPI, mounted with PBS, and covered with a coverslip.

3.3 Laser Scanning Microscopy

Blue, green, or red confocal images were collected using a confocal laser scanning microscopy system Zeiss LSM510 (Jena, Germany) equipped on an inverted epifluorescence microscope Zeiss Axiovert100M (Jena, Germany). A water immersion objective lens Zeiss C-Apochromat 63×/1.2 w corr. (Jena, Germany) was used. UV or Ar laser was tuned at 364 and 543 nm to excite DAPI and Cy3 fluorophores, respectively. The fluorescence signals from each fluorophore were recorded separately in multiple scans to minimize optical crosstalk. Digital images were manipulated using Zeiss LSM510 software (Jena, Germany) and then processed using Adobe Photoshop (San Jose, CA) to make figures.

3.4 Brief Result

Figure 1 shows dynamic conformation changes of chromosome during S-phase tightly coupled with the transition of replication foci distribution in S-PCCs following Cy3-dUTP loading. Cy3-dUTP loading procedure was finished within 10 min followed by 10 min of PCC induction and fixation. Accordingly, only DNA replicated in this short lapse fluorescence. Thus, the observed S-PCCs in the present study reflected the replication stages at most 20 min (10 min for Cy3-dUTP labeling and 10 min for PCC induction) before the cell fixation (*see* Subheading 3.2.1). The different diagnostic appearance of different S-PCCs facilitates the DNA replication stages to be identified precisely. Briefly, clear and precise spatial and temporal distribution of replication foci on chromosomes were clearly seen. (i) In early S-phase, PCCs showed a cloudy spreading mass of thin fibers like a "nebula." Numerous fine granular foci were distributed homogeneously overall on the fibers (Fig. 1c), showing "beads on a string" appearance. (ii) In middle S-phase, typical phase specific "pulverized" PCCs were recognized, and the size of foci increased as the number of foci decreased and distributed unevenly on chromosomes. (Fig. 1f). (iii) In late S-phase, chromosomes showed mostly condensed like mitotic chromosomes. Cy3-dUTP-incorporated regions were recognized as "band arrays" inserted in the condensed chromosome (Fig. 1i, indicated by arrows). The size of foci increased and the number decreased to the level that is easy to score. (iv) In very late S-phase, the number of foci further reduced and predominantly localized at the centromere or telomere regions (Fig. 1l, indicated by arrows). These regions are actually known as satellite heterochromatic DNA regions where DNA replicates at very late S-phase [43].

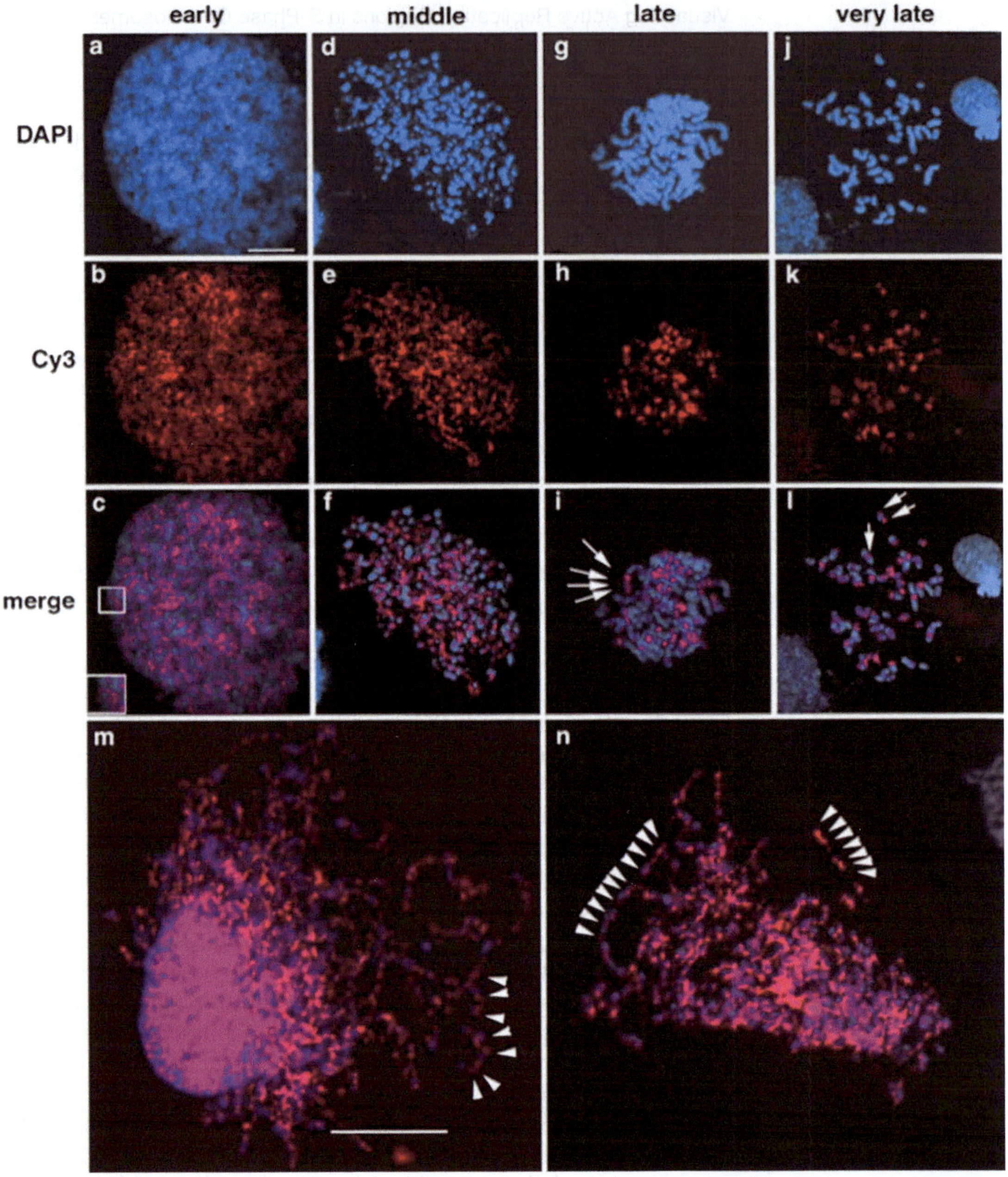

Fig. 1 (1) DNA replication regions on prematurely condensed chromosomes (PCCs) of different substages of S-phase. Approximately 10 min after Cy3-dUTP loading, cells were condensed prematurely using 50 nM of calyculin A [23]. From the left to right column, (A,B,C) early S-phase PCCs, (D,E,F) middle S-PCCs, (G,H,I) late S-PCCs, and (J,K,L) very late S-PCCs. (A,D,G,J) DAPI counterstained DNA, (B,E,H,K) Cy3-dUTP labelled DNA replication region, and (C,F,I,L) merged image of DAPI and Cy3. (L) Centromeric region (arrow) or telomeric region (arrowhead) replicates in very late S-phase are indicated. (I,L) Late S- and very late S-PCCs already condensed like as mitotic chromosomes, but these PCCs were actually S-phase chromosomes because they incorporated Cy3-dUTP. G2/M chromosomes are easily distinguished from late or very late S-phase chromosomes as G2/M chromosomes do not incorporate Cy3-dUTP (data not shown). Inset in (C) is higher magnification of the boxed portion. Scale bar, 10 μm. (2) DNA replication regions seen on prominent fiber of PCCs. (M) Early-S-phase (918 foci scored) and (N) middle S-phase (707 foci scored). Replication foci are clearly seen as "beads on a string" structure; some of these are indicated by arrowhead. Scale bar, 10 μm

Table 1
The number, size, and spacing of replication foci of different subphases of S

Subphase of S	Early	Middle	Late	Very late
Number of foci (max~min)[a]	1046 ± 187 (1396 ~ 758)	678 ± 83 (887 ~ 601)	450 ± 132 (607 ~ 237)	153 ± 42 (245 ~ 98)
Size of foci (max~min) (μm radius)[b,c]	0.35 ± 0.01 (0.08[d]~0.62)	0.52 ± 0.06 (0.1[d]~0.62)	0.98 ± 0.23 (0.24 ~ 1.44)	1.1 ± 0.44 (0.32 ~ 4.1)
Spacing between foci (max~min) (μm)[b,c]	1.56 ± 0.68 (0.64 ~ 4.02)	1.01 ± 0.11 (0.33 ~ 1.86)	N.D.[e]	N.D.[e]

[a]For each subphase, at least 12 spreads were scored, except for very late S (ten spreads to be scored were available)
[b]Measured using the Zeiss LSM510 software
[c]Randomly selected 20 points were measured, and the average and the error were calculated
[d]These data were beyond the resolution of optical microscope and measured on digitized images
[e]The measurement was not done because PCCs at these stages mostly condense as mitotic chromosomes (see Fig. 1); thus measuring the spacing seems less meaningful

Table 1 shows the summaries of number and size of replication foci seen in different subphases of S-phase (i.e., from very early to late S-phase). For more detailed results and following discussion, please see [30].

4 Notes

1. To label the replicating DNA, BrdU penetrates into cell nucleus through cell membrane, so it is very easy to label the replicating DNA [23]. Following immune-chemical reaction to visualize the BrdU-incorporated regions is also established. But immune complex is too large to localize the replication regions in a fine level.
2. Due to the above reason, Cy3-dUTP is more preferable to label replicating DNA as Cy3-dUTP is directly incorporated in DNA and emits bright fluorescence. But Cy3-dUTP does not pass through cell membrane to enter in cells. Therefore, the methods which allow Cy3-dUTP to enter cell through cell membrane are required, such as (1) microinjection method [44] or (2) bead loading method [2].
3. Alternative fluorophore is well-known GFP (green fluorescence protein) which is able to label protein or nucleic acid in living cells [45]. However GFP activity is very sensitive to acidic pH, therefore, resulting in too quenched to detect fluorescence from chromosome spread after fixed with Carnoy's fixative that contain strong acid. I tried to fix chromosomes without the use of acetic acid or decrease the concentration of acetic acid in fixative, but I did not succeed in getting well-spread

chromosome samples to observe with sufficient good quality. So, I finally decided to use Cy3-dUTP to label replicating DNA combined with bead loading method.

4. The cell line used for the Cy3-dUTP labelling by bead loading method is exclusively attached cell line because it is impossible to manipulate bead loading method to the cells that grow in suspension.
5. The size of beads which gives best loading efficiency of Cy3-dUTP into the cells varies and depends on individual cell types. Therefore, pretrial loading experiment using different size of beads should be done to choose the best size of beads.
6. After storing in longer time, esterification reaction will proceed, resulting in production of methyl acetate and water. Consequently, the fixative smells some kinds of aroma, and dropped on a glass slide, fixative does not spread well circularly and water will be remained after the fixative evaporated. In this case, the fixative got old, and new fixative should be prepared.
7. The number of tapping times, strength of tapping, as well as confluency of cells and the beads/cell ratio may influence the loading efficiency of Cy3-dUTP into the cells. So as seen **Note 5**, pretrial loading experiment should be done.
8. All steps should be carried out quick as possible, and prevent the cells from unnecessary fall down the temperature, otherwise active DNA replication will be delay or arrest.
9. Recently developed synthetic transporter may allow Cy3-dUTP to enter into cell more easily and more efficiently without the use of bead loading method or microinjection [46].

References

1. Zink D, Cremer T, Saffrich R, Fischer R, Trendelenburg MF, Ansorge W, Stelzer EH (1998) Structure and dynamics of human interphase chromosome territories in vivo. Hum Genet 102(2):241–251
2. Manders EM, Kimura H, Cook PR (1999) Direct imaging of DNA in living cells reveals the dynamics of chromosome formation. J Cell Biol 144(5):813–821
3. Samaniego R, de la Torre C, Moreno Diaz de la Espina S (2002) Dynamics of replication foci and nuclear matrix during S phase in Allium cepa L. cells. Planta 215(2):195–204
4. Fakan S, Hancock R (1974) Localization of newly-synthesized DNA in a mammalian cell as visualized by high resolution autoradiography. Exp Cell Res 83(1):95–102
5. Edenberg HJ, Huberman JA (1975) Eukaryotic chromosome replication. Annu Rev Genet 9:245–284
6. Hand R (1978) Eucaryotic DNA: organization of the genome for replication. Cell 15(2): 317–325
7. Nakayasu H, Berezney R (1989) Mapping replicational sites in the eucaryotic cell nucleus. J Cell Biol 108(1):1–11
8. Mills AD, Blow JJ, White JG, Amos WB, Wilcock D, Laskey RA (1989) Replication occurs at discrete foci spaced throughout nuclei replicating in vitro. J Cell Sci 94(Pt 3): 471–477
9. Nakamura H, Morita T, Sato C (1986) Structural organizations of replicon domains during

DNA synthetic phase in the mammalian nucleus. Exp Cell Res 165(2):291–297
10. Gotoh E, Durante M (2006) Chromosome condensation outside of mitosis: mechanisms and new tools. J Cell Physiol 209(2): 297–304. and Cover Picture. https://doi.org/10.1002/jcp.20720
11. Johnson RT, Rao PN, Hughes HD (1970) Mammalian cell fusion. 3. A HeLa cell inducer of premature chromosome condensation active in cells from a variety of animal species. J Cell Physiol 76(2):151–157
12. Sperling K, Rao PN (1974) The phenomenon of premature chromosome condensation: its relevance to basic and applied research. Humangenetik 23(4):235–258
13. Hittelman WN, Rao PN (1976) Premature chromosome condensation. Conformational changes of chromatin associated with phytohemagglutinin stimulation of peripheral lymphocytes. Exp Cell Res 100(2):219–222
14. Rao PN, Wilson B, Puck TT (1977) Premature chromosome condensation and cell cycle analysis. J Cell Physiol 91(1):131–141
15. Hanks SK, Rao PN (1980) Initiation of DNA synthesis in the prematurely condensed chromosomes of G1 cells. J Cell Biol 87(1): 285–291. https://doi.org/10.1083/jcb.87.1.285
16. Lau YF, Arrighi FE (1981) Studies of mammalian chromosome replication. II. Evidence for the existence of defined chromosome replicating units. Chromosoma 83(5):721–741
17. Hittelman WN, Pollard M (1984) Visualization of chromatin events associated with repair of ultraviolet light-induced damage by premature chromosome condensation. Carcinogenesis 5(10):1277–1285. https://doi.org/10.1093/carcin/5.10.1277
18. Mullinger AM, Johnson RT (1980) Packing DNA into chromosomes. J Cell Sci 46:61–86
19. Mullinger AM, Johnson RT (1983) Units of chromosome replication and packing. J Cell Sci 64:179–193
20. Hittelman WN (1986) The technique of premature chromosome condensation to study the leukemic process: review and speculations. Crit Rev Oncol Hematol 6(2):147–221. https://doi.org/10.1016/s1040-8428(86)80020-8
21. Johnson RT, Rao PN (1970) Mammalian cell fusion: induction of premature chromosome condensation in interphase nuclei. Nature 226(247):717–722
22. Pantelias GE, Maillie HD (1983) A simple method for premature chromosome condensation induction in primary human and rodent cells using polyethylene glycol. Somatic Cell Genet 9(5):533–547. https://doi.org/10.1007/bf01574257
23. Gotoh E, Asakawa Y, Kosaka H (1995) Inhibition of protein serine/threonine phophatases directly induces premature chromosome condensation in mammalian somatic cells. Biomed Res 16(1):63–68
24. Gotoh E, Asakawa Y (1996) Detection and evaluation of chromosomal aberrations induced by high doses of gamma-irradiation using immunogold-silver painting of prematurely condensed chromosomes. Int J Radiat Biol 70(5):517–520. https://doi.org/10.1080/095530096144716
25. Asakawa Y, Gotoh E (1997) A method for detecting sister chromatid exchanges using prematurely condensed chromosomes and immunogold-silver staining. Mutagenesis 12(3):175–177. https://doi.org/10.1093/mutage/12.3.175
26. Durante M, Furusawa Y, Gotoh E (1998) A simple method for simultaneous interphase-metaphase chromosome analysis in biodosimetry. Int J Radiat Biol 74(4):457–462. https://doi.org/10.1080/095530098141320
27. Terzoudi GI, Malik SI, Pantelias GE, Margaritis K, Manola K, Makropoulos W (2003) A new cytogenetic approach for the evaluation of mutagenic potential of chemicals that induce cell cycle arrest in the G2 phase. Mutagenesis 18(6):539–543
28. Srebniak MI, Trapp GG, Wawrzkiewicz AK, Kazmierczak W, Wiczkowski AK (2005) The usefulness of calyculin A for cytogenetic prenatal diagnosis. J Histochem Cytochem 53(3): 391–394
29. El Achkar E, Gerbault-Seureau M, Muleris M, Dutrillaux B, Debatisse M (2005) Premature condensation induces breaks at the interface of early and late replicating chromosome bands bearing common fragile sites. Proc Natl Acad Sci U S A 102(50):18069–18074. https://doi.org/10.1073/pnas.0506497102
30. Gotoh E (2007) Visualizing the dynamics of chromosome structure formation coupled with DNA replication. Chromosoma 116(5): 453–462. https://doi.org/10.1007/s00412-007-0109-5
31. Lamadrid AI, González JE, García O, Voisin P, Roy L (2011) Prematurely condensed chromosome rings after neutron irradiation of human lymphocytes. J Radiat Res 52(4):531–535. https://doi.org/10.1269/jrr.10096
32. Miura T, Blakely WF (2011) Optimization of calyculin A-induced premature chromosome condensation assay for chromosome aberration

studies. Cytometry A 79(12):1016–1022. https://doi.org/10.1002/cyto.a.21154
33. Ono T, Yamashita D, Hirano T (2013) Condensin II initiates sister chromatid resolution during S phase. J Cell Biol 200(4):429–441. https://doi.org/10.1083/jcb.201208008
34. González JE, Romero I, Gregoire E, Martin C, Lamadrid AI, Voisin P, Barquinero JF, García O (2014) Biodosimetry estimation using the ratio of the longest:shortest length in the premature chromosome condensation (PCC) method applying autocapture and automatic image analysis. J Radiat Res 55(5):862–865. https://doi.org/10.1093/jrr/rru030
35. Sun M, Moquet J, Barnard S, Lloyd D, Ainsbury E (2020) A simplified calyculin A-induced premature chromosome condensation (PCC) protocol for the biodosimetric analysis of high-dose exposure to gamma radiation. Radiat Res 193(6):560–568. https://doi.org/10.1667/rr15538.1
36. Meenakshi C, Venkatachalam P, Chandrasekaran S, Venkatraman B (2021) Construction of dose response curve for 6MV LINAC X-rays using premature chromosome condensation assay for radiation dosimetry. Appl Radiat Isot:109729. https://doi.org/10.1016/j.apradiso.2021.109729
37. McNeil PL, Warder E (1987) Glass beads load macromolecules into living cells. J Cell Sci 88 (Pt 5):669–678
38. Ito S, Gotoh E, Ozawa S, Yanagi K (2002) Epstein-Barr virus nuclear antigen-1 is highly colocalized with interphase chromatin and its newly replicated regions in particular. J Gen Virol 83(Pt 10):2377–2383
39. Johnson RT, Gotoh E, Mullinger AM, Ryan AJ, Shiloh Y, Ziv Y, Squires S (1999) Targeting double-strand breaks to replicating DNA identifies a subpathway of DSB repair that is defective in ataxia-telangiectasia cells. Biochem Biophys Res Commun 261(2):317–325. https://doi.org/10.1006/bbrc.1999.1024
40. Gotoh E, Kawata T, Durante M (1999) Chromatid break rejoining and exchange aberration formation following gamma-ray exposure: analysis in G2 human fibroblasts by chemically induced premature chromosome condensation. Int J Radiat Biol 75(9):1129–1135. https://doi.org/10.1080/095530099139601
41. Gotoh E, Tanno Y (2005) Simple biodosimetry method for cases of high-dose radiation exposure using the ratio of the longest/shortest length of Giemsa-stained drug-induced prematurely condensed chromosomes (PCC). Int J Radiat Biol 81(5):379–385. https://doi.org/10.1080/09553000500147667
42. Gotoh E, Tanno Y, Takakura K (2005) Simple biodosimetry method for use in cases of high-dose radiation exposure that scores the chromosome number of Giemsa-stained drug-induced prematurely condensed chromosomes (PCC). Int J Radiat Biol 81(1):33–40. https://doi.org/10.1080/09553000500092319
43. O'Keefe RT, Henderson SC, Spector DL (1992) Dynamic organization of DNA replication in mammalian cell nuclei: spatially and temporally defined replication of chromosome-specific alpha-satellite DNA sequences. J Cell Biol 116(5):1095–1110
44. Fauth C, Zink D (2001) Live-cell microscopy of single nuclear chromosomes and genome compartments: evaluation of labeling procedure and imaging conditions. Cytometry 45(3):214–224. https://doi.org/10.1002/1097-0320(20011101)45:3<214::aid-cyto1165>3.0.co;2-b
45. Dovrat D, Dahan D, Sherman S, Tsirkas I, Elia N, Aharoni A (2018) A live-cell imaging approach for measuring DNA replication rates. Cell Rep 24(1):252–258. https://doi.org/10.1016/j.celrep.2018.06.018
46. Kužmová E, Zawada Z, Navrátil M, Günterová J, Kraus T (2021) Flow cytometric determination of cell cycle progression via direct labeling of replicated DNA. Anal Biochem 614:114002. https://doi.org/10.1016/j.ab.2020.114002

Chapter 15

Hi-C Analysis to Identify Genome-Wide Chromatin Structural Aberration in Cancer

Atsushi Okabe and Atsushi Kaneda

Abstract

Hi-C is a method that analyzes genome-wide chromatin structure using next-generation sequencer. Chromatin structure is crucial for regulating transcription or replication, and Hi-C has revealed the hierarchical chromatin structures, such as loop, domain, and compartment structures. Aberrant alteration of these structures causes disease, and a number of structural aberrations in cancer cells have been reported recently. Besides, Hi-C can identify chromosome rearrangements that frequently occurred in cancer. Therefore, Hi-C is a powerful technique to analyze epigenomic and genomic aberrations in tumorigenesis. Here we will introduce the basic protocol of Hi-C in experimental and analytical aspects.

Key words Hi-C, Chromosome conformation capture, Chromatin structure, Genome architecture, Chromosome rearrangement, Cancer

1 Introduction

Chromatin structures function essentially in gene expression and maintenance of genomic information; thus, that has been studied for many years [1]. Chromosome conformation capture (3C) is the first method to quantify a chromatin interaction between two regions and elucidated looping structures such as interactions between enhancers and target promoters or interactions between insulators [2]. 3C-based methods have been improved by next-generation sequencers (NGS) such as 4C [3], 5C, ChIA-PET [4], Hi-ChIP [5], and Hi-C [6]. Hi-C is the technique to analyze genome-wide chromatin structure, and it has revealed submega-base-scale chromatin domain structure termed topologically associating domains (TADs) [7, 8] and two distinct chromatin compartment states – A (active) and B (inactive) [6].

These hierarchical chromatin structures play essential roles in regulating cell-type-specific gene expression or expression changes during differentiation [9, 10]. Therefore, their aberrations cause

Eisuke Gotoh (ed.), *Chromosome Analysis: Methods and Protocols*, Methods in Molecular Biology, vol. 2519,
https://doi.org/10.1007/978-1-0716-2433-3_15,

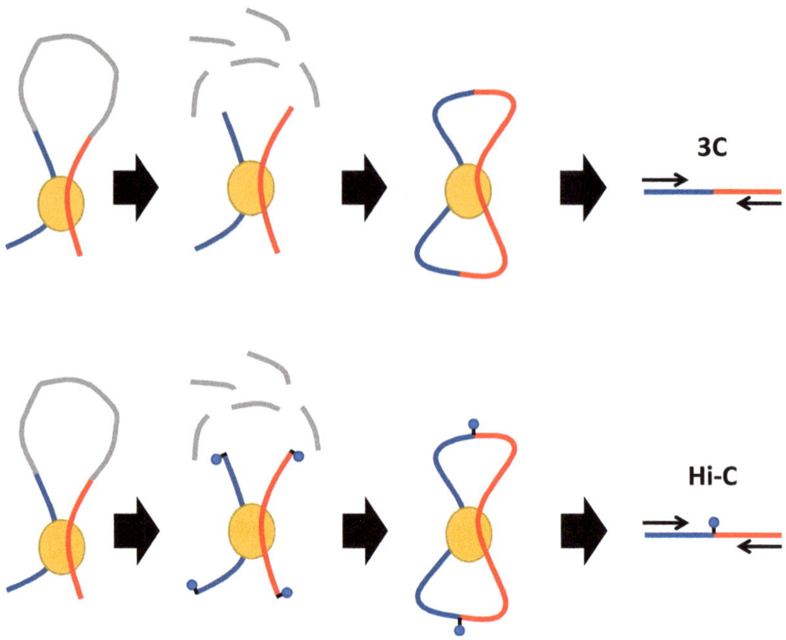

Fig. 1 Overview of chromosome conformation capture-based methods

many diseases, including cancer. Hi-C has been performed using cancer cell lines and patient-derived cancer tissue and identified structural abnormalities such as TAD boundary disruptions and compartment shifts [11–17]. In addition to 3D chromatin structure, Hi-C can clarify many chromosome rearrangements, including copy number variation and translocation, frequently observed in cancer cells [18–20]. Hi-C and its derivative analysis are powerful techniques to identify genomic and epigenomic aberrations in cancer.

In the 3C-based method, restriction enzyme fragmented formaldehyde-fixed chromatin is followed by proximity ligation of DNA end. These steps lead to a unique hybrid DNA molecule that contains two interacting genomic regions. After purification of products, this hybrid DNA molecule is quantified by real-time PCR or NGS (Fig. 1). The first Hi-C method is called "dilution Hi-C." This method dilutes fragmented chromatin in a ligation buffer to avoid ligation between different chromatin complexes. However, inter-chromatin ligation products were generated, and these unintentional products cause noises [6]. The recently developed Hi-C method is called "in situ Hi-C," where fragmented chromatin is ligated in the nucleus without disrupting the nuclear membrane [21, 22]. This development can prevent random ligation between different chromatin complexes, thus reducing noises [23]. In addition to experimental development, many analysis methods and programs using Hi-C data have been progressed [24, 25].

In this chapter, we provide a step-by-step protocol of the in situ Hi-C method and data analysis. Then we briefly introduce analysis programs that have been reported to analyze chromatin structural changes.

2 Materials

2.1 Reagents

1. 37% Formaldehyde.
2. 2.5 M Glycine.
3. PBS.
4. Roche cOmplete Protease Inhibitor Cocktail (Roche, 11697498001).
5. Hi-C Lysis buffer (10 mM Tris-HCl pH 8.0, 10 mM NaCl, 0.2% Igepal CA630).
6. Reverse-crosslinking buffer (10 mM Tris-HCl pH 8, 0.5 M NaCl, 1% SDS solution).
7. 1× Tween Washing Buffer (1× TWB: 5 mM Tris-HCl pH 7.5, 0.5 mM EDTA, 1 M NaCl, 0.05% Tween 20).
8. 2× Binding Buffer (2× BB: 10 mM Tris-HCl pH 7.5, 1 mM EDTA, 2 M NaCl).
9. Restriction enzyme MboI (NEB, R0147).
10. dCTP.
11. dGTP.
12. dTTP.
13. biotin-14-dATP (Life Technologies, 19524-016).
14. 5 U/μl DNA Polymerase I, Large (Klenow) Fragment (NEB, M0210).
15. 5 U/μl T4 DNA Ligase (Roche, 10799009001).
16. SDS.
17. Triton X-100.
18. BSA.
19. Proteinase K (NEB, P8102).
20. AMPure XP beads (Beckman Coulter, A63881).
21. Dynabeads MyOne Streptavidin T1 beads (Life Technologies, 65602).
22. KAPA HyperPrep Kits (Roche, KK8500).
23. KAPA HiFi HS ReadyMix (Roche, KK2601).

2.2 Equipment

1. ThermoMixer (Eppendorf).
2. Centrifuge.

3. Tube rotator.
4. Magnetic separation rack (Thermo Fisher Scientific).
5. Covaris S220 (Covaris, Woburn, MA).
6. Agilent TapeStation (Agilent).
7. Real-time PCR system.
8. PCR thermal cycler.
9. Promega Quantus Fluorometer (Promega).
10. Illumina next-generation sequencer (Illumina).

3 Methods

3.1 Crosslinking

1. Culture cells with 20 ml of culturing medium using 15 cm dish to about 80–90% confluence (*see* **Note 1**).
2. Add 540 μl of 37% formaldehyde solution to 20 ml medium in 15 cm dish (a final concentration of 1%, v/v), and incubate at room temperature for 10 min with shaking.
3. Add 1.8 ml of 2.5 M glycine solution to a final concentration of 0.2 M to quench the reaction, and incubate at room temperature for 5 min with shaking.
4. Wash cells by cold PBS three times. Add 5 ml cold PBS, collect cells using a cell scraper, and transfer to a new 15 ml tube.
5. Centrifuge for 5 min at 300 × G at 4 °C and discard the supernatant.
6. Flash-freeze cell pellets in liquid nitrogen.
7. Either proceed to the rest of the protocol or store cell pellets at −80 °C.

3.2 Digestion

1. Mix ice-cold Hi-C lysis buffer (10 mM Tris-HCl pH 8.0, 10 mM NaCl, 0.2% Igepal CA630) with Roche cOmplete Protease Inhibitor Cocktail (Roche, 11697498001).
2. Add 500 μl lysis buffer to 5 × 10^6 crosslinked cells, resuspend well by pipetting, and incubate cell suspension at 4 °C for 30 min with rotation (*see* **Note 2**).
3. Centrifuge at 2500 × G for 5 min and discard the supernatant.
4. Wash pelleted nuclei once with 500 μl of ice-cold Hi-C lysis buffer.
5. Gently resuspend pellet in 100 μl of 0.5% sodium dodecyl sulfate (SDS) and incubate at 62 °C for 10 min.
6. After heating is over, add 285 μl of water and 50 μl of 10% Triton X-100 (Sigma, 93443) to quench the SDS. Mix well, avoiding excessive foaming, and incubate at 37 °C for 15 min with rotation.

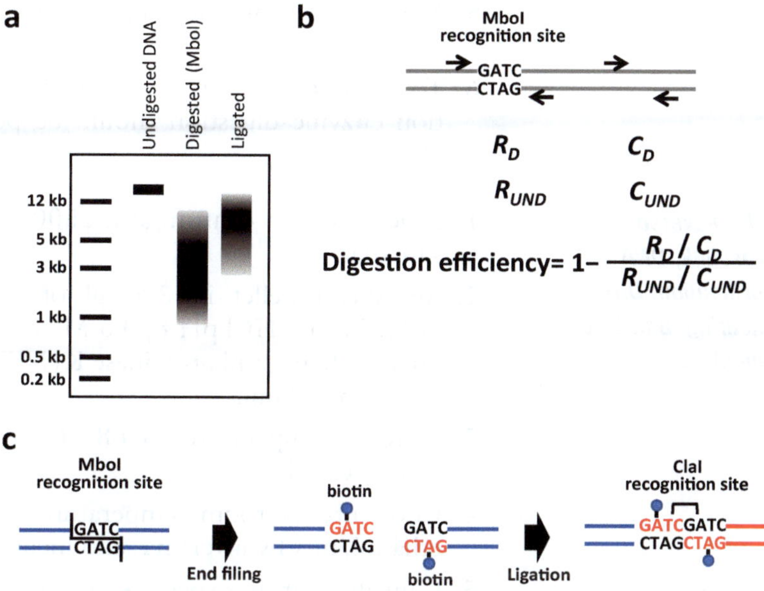

Fig. 2 Quality check of digestion and ligation. (**a**) Schematic representation of electrophoresis image for digestion check. The DNA from the undigested control should show a high molecular weight band (>12 kb). The digested DNA with MboI should show smear band less than 5 kb. The ligated DNA should show a clear upwards shift in molecular weight. (**b**) Primer design to calculate digestion efficiency and the calculation formula. (**c**) The sequence of ligated sample at restriction digested sites after end filling and ligation. Ligation of the blunted ends creates a recognition site of a new restriction enzyme, ClaI

7. Add 50 μl of 10× NEBuffer 2 and 8 μl of 25 U/μl MboI (total 200 U) restriction enzyme (NEB, R0147) and digest chromatin overnight at 37 °C with rotation (*see* **Note 3**).
8. Check digestion efficiency by electrophoresis or real-time PCR (Fig. 2a, b; *see* **Note 4**).

3.3 Marking of DNA Ends and Proximity Ligation

1. Incubate at 62 °C for 20 min to inactivate restriction enzyme without shaking or rotation and then cool to room temperature.
2. To fill in the restriction fragment overhangs and mark the DNA ends with biotin, add 50 μl of fill-in master mix (37.5 μl of 0.4 mM biotin-14-dATP (Life Technologies, 19524-016), 1.5 μl of 10 mM dCTP, 1.5 μl of 10 mM dGTP, 1.5 μl of 10 mM dTTP, 10 μl of 5 U/μl DNA Polymerase I, Large (Klenow) Fragment (NEB, M0210)).
3. Mix by pipetting and incubate at 37 °C for 1 h with rotation.
4. Add 948 μl of ligation master mix (654 μl of water, 150 μl of 10× T4 DNA ligase buffer, 125 μl of 10% Triton X-100, 15 μl of 10 mg/ml bovine serum albumin, 4 μl of 5 U/μl T4 DNA Ligase (Roche, 10799009001)).

5. Mix by inverting and incubate at room temperature for 4 h with slow rotation.
6. To check the ligation efficiency, PCR amplification and restriction enzyme digestion should be performed (Fig. 2c, d; *see* **Note 5**).

3.4 Reverse Crosslink, DNA Purification, DNA Shearing, and Size Selection

1. Pellet nuclei by centrifuge at 2500 × G for 5 min. Discard the supernatant.
2. Resuspend pellet in 300 μl of reverse-crosslinking buffer (10 mM Tris-HCl pH 8, 0.5 M NaCl, 1% SDS solution). Add 50 μl of 20 mg/ml proteinase K (NEB, P8102) and incubate at 55 °C for 30 min.
3. Increase temperature to 68 °C and incubate overnight to reverse crosslinks.
4. Cool tube at room temperature. Add 2× volumes of pure ethanol, mix by inverting, and incubate at −80 °C for 15 min.
5. Centrifuge at max speed, 4 °C for 30 min. Keep the tubes on ice after spinning and carefully remove the supernatant by pipetting.
6. Wash the DNA pellet with 800 μl of 70% ethanol. Centrifuge at max speed for 5 min.
7. Remove all supernatant and wash the pellet once more with 800 μl of 70% ethanol.
8. Dissolve the pellet in 130 μl of 10 mM Tris-HCl pH 8 and incubate at 37 °C for 15 min to fully dissolve the DNA.
9. To make the biotinylated DNA suitable for high-throughput sequencing using Illumina sequencers, shear to a size of 300–500 bp using Covaris S220 (Covaris, Woburn, MA) at the following parameters:
   - (Volume of library: 130 μl in a Covaris microtube; fill level: 10; duty cycle: 15; PIP: 500; cycles/burst: 200; time: 58 s).
10. Transfer sheared DNA to a fresh 1.5 ml tube.
11. Wash the Covaris vial with 70 μl of water and add to the sample, bringing the total reaction volume to 200 μl.
12. Run a 1 μl of sample on Agilent TapeStation to verify successful shearing.
13. Add exactly 110 μl (0.55× volumes) of AMPure XP beads to the reaction. Mix well by pipetting and incubate at room temperature for 5 min.
14. Separate on a magnet. Transfer the clear solution to a fresh tube, avoiding any beads. (The supernatant will contain fragments shorter than 500 bp.)
15. Add precisely 30 μl of fresh AMPure XP beads to the solution. Mix by pipetting and incubate at room temperature for 5 min.

16. Separate on a magnet and keep the beads. (Fragments in the range of 300–500 bp will be retained on the beads.)
17. Discard the supernatant.
18. Keeping the beads on the magnet, wash twice with 700 μl of 70% ethanol without mixing.
19. To elute DNA, add 300 μl of 10 mM Tris-HCl (pH 7.4), gently mix by pipetting, incubate at room temperature for 5 min, separate on a magnet, and transfer the solution to a fresh 1.5 ml tube.
20. Quantify DNA by Quantus Fluorometer (Promega) and run 1 μL on Agilent TapeStation to verify successful size selection.

3.5 Biotin Pull-Down and Preparation for Illumina Sequencing

1. Perform all the following steps in low-bind tubes.
2. Prepare for biotin pull-down by washing 150 μl of 10 mg/ml Dynabeads MyOne Streptavidin T1 beads (Life Technologies, 65602) with 400 μl of 1× Tween Washing Buffer (1× TWB: 5 mM Tris-HCl pH 7.5, 0.5 mM EDTA, 1 M NaCl, 0.05% Tween 20).
3. Separate on a magnet and discard the solution.
4. Resuspend the beads in 300 μl of 2× Binding Buffer (2 × BB: 10 mM Tris-HCl pH 7.5, 1 mM EDTA, 2 M NaCl) and add to the reaction.
5. Incubate at room temperature for 15 min with rotation to bind biotinylated DNA to the streptavidin beads.
6. Separate on a magnet and discard the solution.
7. Wash the beads by adding 600 μl of 1× TWB and transferring the mixture to a new tube.
8. Heat the tubes on a ThermoMixer at 55 °C for 2 min with mixing.
9. Reclaim the beads using a magnet. Discard supernatant.
10. Repeat wash.
11. Resuspend beads in 60 ul A-tailing mix (KAPA HyperPrep Kits; 50 μl of DDW, 7 μl of End Repair & A-tailing buffer, 3 μl of End Repair & A-tailing enzyme mix).
12. Incubate at 20 °C for 30 min, and then incubate at 65 °C for 30 min.
13. Add 40 μl of ligation master mix (KAPA HyperPrep Kits; 5 μl of DDW, 5 μl of Adaptor stock, 30 μl of ligation buffer, 10 μl of DNA ligase), and incubate at 20 °C for 15 min.
14. Separate on a magnet and discard the solution.
15. Wash the beads by adding 600 μl of 1× TWB and incubate the tubes on a ThermoMixer at 55 °C for 2 min with mixing.

16. Reclaim the beads using a magnet. Discard supernatant.
17. Wash beads by adding 100 μl of 10 mM Tris-HCl (pH 7.4). Reclaim beads and discard the solution.
18. Resuspend in 50 μl of 10 mM Tris-HCl (pH 7.4).
19. Incubate at 98 °C for 10 min with shaking to remove biotinylated DNA from beads.
20. Separate on a magnet and transfer the solution to a new 1.5 ml tube.

3.6 Final Amplification and Purification

1. To estimate cycle number, perform real-time PCR using 1 μL of the sample with KAPA HiFi HS ReadyMix (5 μl of KAPA 2× PCR master mix, 1 μl of library amplification primer mix, 1 μl of X10 SYBR Green, 2 μl of DDW, 1 μL of sample template). The following PCR program was performed: 72 °C for 5 min, 98 °C for 1 min, then cycle at 98 °C for 15 s, 63 °C for 30 s, and 72 °C for 1 min (*see* **Note 6**).
2. After the cycle number is determined, PCR final amplification will be performed (25 μl of KAPA HiFi HS ReadyMix, 5 μl of library amplification primer mix, 2 μL of DDW, 18 μl of sample template).
3. After amplification is complete, transfer the sample to a new 1.5 ml tube.
4. Add 90 μl of AMPure XP beads to the PCR reaction (1.8× volumes). Mix by pipetting and incubate at room temperature for 5 min.
5. Separate on a magnet and remove the clear solution.
6. Keeping the beads on the magnet, wash twice with 700 μl of 70% ethanol without mixing.
7. Remove ethanol completely.
8. Resuspend in 20 μl of 10 mM Tris-HCl (pH 7.4) to elute DNA.
9. Mix by pipetting and incubate at room temperature for 5 min.
10. Separate on a magnet, and transfer the solution to a fresh labeled tube. The result is a final in situ Hi-C library ready to be quantified.
11. Sequence Hi-C libraries using Illumina platform 75 bp or 150 bp paired-end sequencing.

3.7 Data Analysis

1. Read alignments and matrix generation

   Hi-C paired-end reads are aligned to a reference genome, and valid interaction pairs are identified using read orientation (inward, outward, or same direction). Valid interaction read pairs are used to make interaction matrices at a range of resolutions. Binned interaction matrices are corrected for intrinsic

Table 1
Program for Hi-C data analysis

Program	Align reads and filter invalid reads	Make visualization file	Calculate compartment	Calculate TAD	Calculate loop	Calculate chromosomal rearrangement
HiC-Pro	✓					
Juicer	✓	✓				
HiCExplorer	✓	✓	✓	✓	✓	
HOMER			✓	✓	✓	
Juicer tools			✓	✓	✓	
Fit-HiC					✓	
HiNT						✓
calCB						✓
HiCnv and HiCtrans						✓

biases such as read mappability, restriction site density, and GC content. There are several public pipelines for processing Hi-C data, e.g., HiCPro [26] or Juicer [27] (Table 1) (*see* **Note 7**).

2. Visualization of Hi-C data

   Several tools to explore Hi-C contact map data have been developed. Contact matrix files, such as hic format or mcool format, are used for visualization (Fig. 3a). Juicebox [28] can show 2D contact matrices interactively with a desktop application and a web browser. Other tools, HiGlass [29] and the WashU Epigenome Browser [30], are web browsers to show contact matrices. These Hi-C browsers can display 1D ChIP-seq data and 2D loop or domain structures together with Hi-C contact matrices.

3. Compartment detection

   After Pearson correlation of the Hi-C contact matrices, a "checker-board"-like pattern is observed (Fig. 3b). This pattern indicates that chromosomes are classified into two genomic regions, and the same type tends to interact more frequently than the other type. Principal component analysis (PCA) is performed to identify these two types of genomic regions, called compartments. The active "A" compartments are associated with euchromatic regions, whereas the inactive "B" compartments are associated with heterochromatic regions. Several tools to identify compartments have been reported, such as HOMER [31], HiCExplorer [32], Juicer tools [27], etc. (Table 1). After PCA analysis, the PC1 value

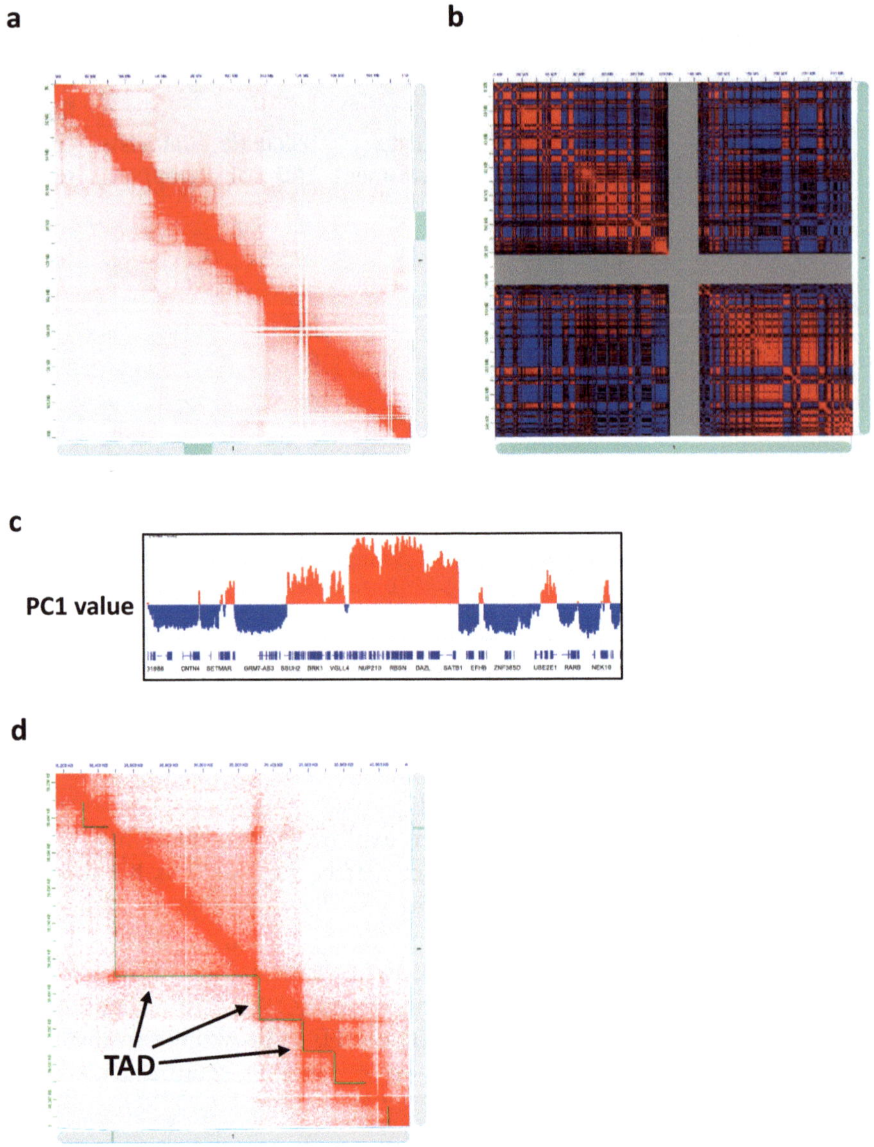

Fig. 3 Analysis and visualization of Hi-C data. (**a**) A Hi-C interaction from chromosome 1. (**b**) Hi-C data from chromosome 1. (**c**) The representative genome browser image of PC1 value at chromosome 1. The regions with PC1 value >0 are A compartment, and the regions with PC1 value <0 are B compartment. (**d**) Calculated TAD. A Hi-C interaction and calculated TADs are shown by Juicebox. Green triangles are TADs

is calculated at each genomic region. The regions with PC1 value >0 are A compartment, and the regions with PC1 value <0 are B compartment (Fig. 3c).

4. TAD detection

Hi-C heatmaps show isolated triangle structures, and they indicate that genomic regions are separated into highly self-interacting domains, called TADs (Fig. 3d). The locations of

TADs can be calculated from binning interaction matrices at 40 kb or less. Several computational approaches to identify TAD boundaries were developed, such as the directionality index or an insulation score algorithm. Tools for TAD calling, such as HOMER [31], Arrowhead in Juicer tools [22], and HiCExplorer [32], have been reported (Table 1).

5. Interaction detection

   To calculate the significant interaction between two loci, interaction matrix data binning in a 10 kb resolution or higher is required. HOMER [31], Fit-Hi-C [33], and HiCCUPS in Juicer tools [22] can be used to calculate interactions (Table 1).

6. Identifying chromosomal rearrangement

   Chromosomal rearrangements are frequently observed in cancer cells, and Hi-C data can also identify rearrangements, including copy number variation, chromosome loss, and translocation. HiNT [34], calCB [35], HiCnv, and HiCtrans [36] can be used to identify chromosomal rearrangement from Hi-C data (Table 1).

4 Notes

1. The number of cells should be counted to estimate the appropriate volume of the restriction enzyme. For adherent cells, an extra culturing dish for cell counting should be prepared.
2. Sometimes cells are aggregated in a lysis buffer. In such a case, aggregated calls should be mixed well by pipetting.
3. Fewer restriction enzymes can be used for the lower number of cells. For example, 100 U of restriction enzyme is used for one million cells.
4. Digestion efficiency by restriction enzyme is critical to success. The size of the undigested control sample and the digested sample should be verified by electrophoresis (Fig. 2a). A tight band of over 20 kb in size is observed in the electrophoresis image of the undigested sample. After digestion, a smeared band with a size range specific for the applied restriction enzyme is observed. Or calculating the digestion efficiency by real-time PCR is better. After quantifying the genomic regions spanning restriction enzyme sites and a control region that does not span a restriction enzyme site, the rate of the digested sample and the undigested sample is calculated (Fig. 2b). To create standard curve, dilution series of undigested DNA are used. Digestion efficiency should be more than 70%.
5. The efficiency of end filling and ligation is a critical step and should be assessed. The ligation product is amplified by PCR using primers that are designed across a restriction enzyme site.

In the ligation product, MboI recognition sites "GATC" become "GATCGATC" after end-filling and ligation (Fig. 2c). Therefore, another restriction enzyme, ClaI, can cut the PCR product at its recognition site "CG." When ClaI digests the majority of PCR products, the efficiency is enough.

6. Hi-C template generated from 5×10^6 cells is amplified in 6–10 cycles. More than ten cycle PCR amplification leads to lower complexity of the sequencing library and a lower rate of valid interaction reads.
7. A high-quality Hi-C library for mammalian genomes typically has 50–70% of interactions mapping to intrachromosomal interactions, less than 2% dangling ends, less than 1% self-ligated circles, and less than 5% PCR redundant interactions per 400 million reads.

References

1. Dekker J, Belmont AS, Guttman M, Leshyk VO, Lis JT, Lomvardas S, Mirny LA, O'Shea CC, Park PJ, Ren B, Politz JCR, Shendure J, Zhong S (2017) The 4D nucleome project. Nature 549(7671):219–226. https://doi.org/10.1038/nature23884
2. Dekker J, Rippe K, Dekker M, Kleckner N (2002) Capturing chromosome conformation. Science 295(5558):1306–1311. https://doi.org/10.1126/science.1067799
3. Zhao Z, Tavoosidana G, Sjölinder M, Göndör A, Mariano P, Wang S, Kanduri C, Lezcano M, Sandhu KS, Singh U, Pant V, Tiwari V, Kurukuti S, Ohlsson R (2006) Circular chromosome conformation capture (4C) uncovers extensive networks of epigenetically regulated intra- and interchromosomal interactions. Nat Genet 38(11):1341–1347. https://doi.org/10.1038/ng1891
4. Fullwood MJ, Liu MH, Pan YF, Liu J, Xu H, Mohamed YB, Orlov YL, Velkov S, Ho A, Mei PH, Chew EG, Huang PY, Welboren WJ, Han Y, Ooi HS, Ariyaratne PN, Vega VB, Luo Y, Tan PY, Choy PY, Wansa KD, Zhao B, Lim KS, Leow SC, Yow JS, Joseph R, Li H, Desai KV, Thomsen JS, Lee YK, Karuturi RK, Herve T, Bourque G, Stunnenberg HG, Ruan X, Cacheux-Rataboul V, Sung WK, Liu ET, Wei CL, Cheung E, Ruan Y (2009) An oestrogen-receptor-alpha-bound human chromatin interactome. Nature 462(7269):58–64. https://doi.org/10.1038/nature08497
5. Mumbach MR, Satpathy AT, Boyle EA, Dai C, Gowen BG, Cho SW, Nguyen ML, Rubin AJ, Granja JM, Kazane KR, Wei Y, Nguyen T, Greenside PG, Corces MR, Tycko J, Simeonov DR, Suliman N, Li R, Xu J, Flynn RA, Kundaje A, Khavari PA, Marson A, Corn JE, Quertermous T, Greenleaf WJ, Chang HY (2017) Enhancer connectome in primary human cells identifies target genes of disease-associated DNA elements. Nat Genet 49(11):1602–1612. https://doi.org/10.1038/ng.3963
6. Lieberman-Aiden E, van Berkum NL, Williams L, Imakaev M, Ragoczy T, Telling A, Amit I, Lajoie BR, Sabo PJ, Dorschner MO, Sandstrom R, Bernstein B, Bender MA, Groudine M, Gnirke A, Stamatoyannopoulos J, Mirny LA, Lander ES, Dekker J (2009) Comprehensive mapping of long-range interactions reveals folding principles of the human genome. Science 326(5950):289–293. https://doi.org/10.1126/science.1181369
7. Nora EP, Lajoie BR, Schulz EG, Giorgetti L, Okamoto I, Servant N, Piolot T, van Berkum NL, Meisig J, Sedat J, Gribnau J, Barillot E, Blüthgen N, Dekker J, Heard E (2012) Spatial partitioning of the regulatory landscape of the X-inactivation centre. Nature 485(7398):381–385. https://doi.org/10.1038/nature11049
8. Dixon JR, Selvaraj S, Yue F, Kim A, Li Y, Shen Y, Hu M, Liu JS, Ren B (2012) Topological domains in mammalian genomes identified by analysis of chromatin interactions. Nature 485(7398):376–380. https://doi.org/10.1038/nature11082
9. Bonev B, Cavalli G (2016) Organization and function of the 3D genome. Nat Rev Genet 17(12):772. https://doi.org/10.1038/nrg.2016.147

10. Gibcus JH, Dekker J (2013) The hierarchy of the 3D genome. Mol Cell 49(5):773–782. https://doi.org/10.1016/j.molcel.2013.02.011
11. Okabe A, Huang KK, Matsusaka K, Fukuyo M, Xing M, Ong X, Hoshii T, Usui G, Seki M, Mano Y, Rahmutulla B, Kanda T, Suzuki T, Rha SY, Ushiku T, Fukayama M, Tan P, Kaneda A (2020) Cross-species chromatin interactions drive transcriptional rewiring in Epstein-Barr virus-positive gastric adenocarcinoma. Nat Genet 52(9):919–930. https://doi.org/10.1038/s41588-020-0665-7
12. Flavahan WA, Drier Y, Johnstone SE, Hemming ML, Tarjan DR, Hegazi E, Shareef SJ, Javed NM, Raut CP, Eschle BK, Gokhale PC, Hornick JL, Sicinska ET, Demetri GD, Bernstein BE (2019) Altered chromosomal topology drives oncogenic programs in SDH-deficient GISTs. Nature 575(7781): 229–233. https://doi.org/10.1038/s41586-019-1668-3
13. Flavahan WA, Drier Y, Liau BB, Gillespie SM, Venteicher AS, Stemmer-Rachamimov AO, Suvà ML, Bernstein BE (2016) Insulator dysfunction and oncogene activation in IDH mutant gliomas. Nature 529(7584):110–114. https://doi.org/10.1038/nature16490
14. Kloetgen A, Thandapani P, Ntziachristos P, Ghebrechristos Y, Nomikou S, Lazaris C, Chen X, Hu H, Bakogianni S, Wang J, Fu Y, Boccalatte F, Zhong H, Paietta E, Trimarchi T, Zhu Y, Van Vlierberghe P, Inghirami GG, Lionnet T, Aifantis I, Tsirigos A (2020) Three-dimensional chromatin landscapes in T cell acute lymphoblastic leukemia. Nat Genet 52(4):388–400. https://doi.org/10.1038/s41588-020-0602-9
15. Yusufova N, Kloetgen A, Teater M, Osunsade A, Camarillo JM, Chin CR, Doane AS, Venters BJ, Portillo-Ledesma S, Conway J, Phillip JM, Elemento O, Scott DW, Béguelin W, Licht JD, Kelleher NL, Staudt LM, Skoultchi AI, Keogh MC, Apostolou E, Mason CE, Imielinski M, Schlick T, David Y, Tsirigos A, Allis CD, Soshnev AA, Cesarman E, Melnick AM (2021) Histone H1 loss drives lymphoma by disrupting 3D chromatin architecture. Nature 589(7841):299–305. https://doi.org/10.1038/s41586-020-3017-y
16. Johnstone SE, Reyes A, Qi Y, Adriaens C, Hegazi E, Pelka K, Chen JH, Zou LS, Drier Y, Hecht V, Shoresh N, Selig MK, Lareau CA, Iyer S, Nguyen SC, Joyce EF, Hacohen N, Irizarry RA, Zhang B, Aryee MJ, Bernstein BE (2020) Large-scale topological changes restrain malignant progression in colorectal cancer. Cell 182(6):1474–1489.e1423. https://doi.org/10.1016/j.cell.2020.07.030
17. Akdemir KC, Le VT, Chandran S, Li Y, Verhaak RG, Beroukhim R, Campbell PJ, Chin L, Dixon JR, Futreal PA (2020) Disruption of chromatin folding domains by somatic genomic rearrangements in human cancer. Nat Genet 52(3):294–305. https://doi.org/10.1038/s41588-019-0564-y
18. Harewood L, Kishore K, Eldridge MD, Wingett S, Pearson D, Schoenfelder S, Collins VP, Fraser P (2017) Hi-C as a tool for precise detection and characterisation of chromosomal rearrangements and copy number variation in human tumours. Genome Biol 18(1):125. https://doi.org/10.1186/s13059-017-1253-8
19. Dixon JR, Xu J, Dileep V, Zhan Y, Song F, Le VT, Yardımcı GG, Chakraborty A, Bann DV, Wang Y, Clark R, Zhang L, Yang H, Liu T, Iyyanki S, An L, Pool C, Sasaki T, Rivera-Mulia JC, Ozadam H, Lajoie BR, Kaul R, Buckley M, Lee K, Diegel M, Pezic D, Ernst C, Hadjur S, Odom DT, Stamatoyannopoulos JA, Broach JR, Hardison RC, Ay F, Noble WS, Dekker J, Gilbert DM, Yue F (2018) Integrative detection and analysis of structural variation in cancer genomes. Nat Genet 50(10):1388–1398. https://doi.org/10.1038/s41588-018-0195-8
20. Wu P, Li T, Li R, Jia L, Zhu P, Liu Y, Chen Q, Tang D, Yu Y, Li C (2017) 3D genome of multiple myeloma reveals spatial genome disorganization associated with copy number variations. Nat Commun 8(1):1937. https://doi.org/10.1038/s41467-017-01793-w
21. Belaghzal H, Dekker J, Gibcus JH (2017) Hi-C 2.0: An optimized Hi-C procedure for high-resolution genome-wide mapping of chromosome conformation. Methods 123: 56–65. https://doi.org/10.1016/j.ymeth.2017.04.004
22. Rao SS, Huntley MH, Durand NC, Stamenova EK, Bochkov ID, Robinson JT, Sanborn AL, Machol I, Omer AD, Lander ES, Aiden EL (2014) A 3D map of the human genome at kilobase resolution reveals principles of chromatin looping. Cell 159(7):1665–1680. https://doi.org/10.1016/j.cell.2014.11.021
23. Nagano T, Várnai C, Schoenfelder S, Javierre BM, Wingett SW, Fraser P (2015) Comparison of Hi-C results using in-solution versus in-nucleus ligation. Genome Biol 16(1):175. https://doi.org/10.1186/s13059-015-0753-7
24. Lajoie BR, Dekker J, Kaplan N (2015) The Hitchhiker's guide to Hi-C analysis: practical

guidelines. Methods 72:65–75. https://doi.org/10.1016/j.ymeth.2014.10.031

25. Ay F, Noble WS (2015) Analysis methods for studying the 3D architecture of the genome. Genome Biol 16:183. https://doi.org/10.1186/s13059-015-0745-7
26. Servant N, Varoquaux N, Lajoie BR, Viara E, Chen CJ, Vert JP, Heard E, Dekker J, Barillot E (2015) HiC-Pro: an optimized and flexible pipeline for Hi-C data processing. Genome Biol 16:259. https://doi.org/10.1186/s13059-015-0831-x
27. Durand NC, Shamim MS, Machol I, Rao SS, Huntley MH, Lander ES, Aiden EL (2016) Juicer provides a one-click system for analyzing loop-resolution hi-C experiments. Cell Syst 3(1):95–98. https://doi.org/10.1016/j.cels.2016.07.002
28. Durand NC, Robinson JT, Shamim MS, Machol I, Mesirov JP, Lander ES, Aiden EL (2016) Juicebox provides a visualization system for hi-C contact maps with unlimited zoom. Cell Syst 3(1):99–101. https://doi.org/10.1016/j.cels.2015.07.012
29. Kerpedjiev P, Abdennur N, Lekschas F, McCallum C, Dinkla K, Strobelt H, Luber JM, Ouellette SB, Azhir A, Kumar N, Hwang J, Lee S, Alver BH, Pfister H, Mirny LA, Park PJ, Gehlenborg N (2018) HiGlass: web-based visual exploration and analysis of genome interaction maps. Genome Biol 19(1):125. https://doi.org/10.1186/s13059-018-1486-1
30. Zhou X, Lowdon RF, Li D, Lawson HA, Madden PA, Costello JF, Wang T (2013) Exploring long-range genome interactions using the WashU Epigenome Browser. Nat Methods 10(5):375–376. https://doi.org/10.1038/nmeth.2440
31. Heinz S, Benner C, Spann N, Bertolino E, Lin YC, Laslo P, Cheng JX, Murre C, Singh H, Glass CK (2010) Simple combinations of lineage-determining transcription factors prime cis-regulatory elements required for macrophage and B cell identities. Mol Cell 38(4):576–589. https://doi.org/10.1016/j.molcel.2010.05.004
32. Wolff J, Bhardwaj V, Nothjunge S, Richard G, Renschler G, Gilsbach R, Manke T, Backofen R, Ramírez F, Grüning BA (2018) Galaxy HiCExplorer: a web server for reproducible Hi-C data analysis, quality control and visualization. Nucleic Acids Res 46(W1):W11–w16. https://doi.org/10.1093/nar/gky504
33. Ay F, Bailey TL, Noble WS (2014) Statistical confidence estimation for Hi-C data reveals regulatory chromatin contacts. Genome Res 24(6):999–1011. https://doi.org/10.1101/gr.160374.113
34. Wang S, Lee S, Chu C, Jain D, Kerpedjiev P, Nelson GM, Walsh JM, Alver BH, Park PJ (2020) HiNT: a computational method for detecting copy number variations and translocations from Hi-C data. Genome Biol 21(1):73. https://doi.org/10.1186/s13059-020-01986-5
35. Wu HJ, Michor F (2016) A computational strategy to adjust for copy number in tumor Hi-C data. Bioinformatics 32(24):3695–3701. https://doi.org/10.1093/bioinformatics/btw540
36. Chakraborty A, Ay F (2018) Identification of copy number variations and translocations in cancer cells from Hi-C data. Bioinformatics 34(2):338–345. https://doi.org/10.1093/bioinformatics/btx664

Chapter 16

CUT&Tag Using "Stress-Free" Con A-Conjugated Dynabeads®

Yasuhiro Fujiwara and Yuki Okada

Abstract

Epigenome research has employed various methods to identify the genomic location of proteins of interest, such as transcription factors and histone modifications. CUT&Tag is a recently established method used in epigenome research to determine the genomic location of proteins of interest, such as transcription factors and histone modifications. In CUT&Tag method, cells are bound and hold on concanavalin A (con A)-coated magnetic beads, then a Protein-A Tn5 transposase fusion protein cuts the genome and inserts adapter sequences nearby the target protein. Here we describe the updated CUT&Tag procedure using "home-made" con A-conjugated magnetic beads. This method is free of poor suspendability and severe aggregation, hence providing improved sensitivity.

Key words CUT&Tag, CUT&RUN, Concanavalin A, Dynabeads, Epigenome, Epigenetics

1 Introduction

Gene transcription is regulated by regulatory factors, including transcription factors (TFs) and histone modifications, which do not involve the alteration of DNA sequence. The study of transcriptional regulation by these factors is called epigenetics, and understanding of these factors is essential for epigenetics research. Chromatin immunoprecipitation with sequencing (ChIP-seq) method has been a primary method to detect the genomic location of epigenetics factors. More recently, two newly invented methods were reported: cleavage under targets and release using nuclease (CUT&RUN) and cleavage under targets and tagmentation (CUT&Tag), the former uses antibody and a Protein A-Micrococcal Nuclease (pA/MNase) fusion protein [1, 2], and the latter uses a Protein A-Tn5 transposase (pA/Tn5) fusion protein on unfixed cells [3, 4], respectively. In many cases, CUT&RUN and CUT&Tag methods appear to be superior to ChIP-seq in the advantages of a mitigated number of cells and

Eisuke Gotoh (ed.), *Chromosome Analysis: Methods and Protocols*, Methods in Molecular Biology, vol. 2519,
https://doi.org/10.1007/978-1-0716-2433-3_16,

sequence reads required, owing to that unlike ChIP-seq, CUT&RUN and CUT&Tag do not require solubilization of the entire genome of cells that indeed generate high background reads.

In both CUT&RUN and CUT&Tag methods, cells are bound to concanavalin A (con A)-coated magnetic beads throughout the process until the DNA extraction step. However, commercially available con A-coated magnetic beads often exhibit unwanted poor suspendability and severe aggregation. These disadvantages might affect the efficiency of enzyme reactions.

Here we describe an updated CUT&Tag protocol using "home-made" concanavalin con A-conjugated magnetic beads. This method is free of poor suspendability and severe aggregation and provides improved sensitivity.

2 Materials

2.1 Preparation of Con A-Conjugated Dynabeads

1. Biotin-conjugated concanavalin A, Type IV (Sigma-Aldrich, C2272).
2. Dynabeads MyOne T1 (Invitrogen, 65601).
3. 1 × PBS (pH 6.8) (*see* **Note 1**).

2.2 Preparation of Tn5 Adapter Complex

1. 3 × Flag-pA-Tn5-Fl (Addgene, #124601).
2. *Escherichia coli* strain BL21-pLysS Singles™ Competent Cells (Novagen, 70236).
3. Ampicillin (100 mg/mL).
4. IPTG (250 mM).
5. 10% neutralized PEI (Polyethylenimine).
6. Chitin resin (NEB, S6651S).
7. Mosaic end-adapter A oligonucleotide (Tn5ME-A): 5′- TCG TCGGCAGCGTCAGATGTGTATAAGAGACAG -3′ (100 μM).
8. Mosaic end-adapter B oligo (Tn5ME-B): 5′-GTCTCGTGGG CTCGGAGATGTGTATAAGAGACAG -3′ (100 μM).
9. Mosaic end-reverse oligonucleotide (Tn5MErev): 5′- [Phos] CTGTCTCTTATACACATCT -3′ (100 μM).
10. HEGX buffer: 20 mM HEPES-KOH (pH 7.2), 0.8 M NaCl, 1 mM EDTA, 10% glycerol, 0.2% Triton X-100.
11. Elution buffer: HEGX buffer containing 100 mM DTT.
12. Dialysis buffer: 100 mM HEPES-KOH (pH 7.2), 100 mM NaCl, 0.2 mM EDTA, 2 mM 2-ME, 0.2% Triton X-100, 20% glycerol.

2.3 CUT&Tag

1. Binding buffer: 20 mM HEPES pH 7.5, 10 mM KCl, 1 mM $CaCl_2$, 1 mM $MnCl_2$ (*see* **Note 2**).
2. Wash buffer: 20 mM HEPES (pH 7.5), 150 mM NaCl, 0.5 mM spermidine, Complete Protease Inhibitor Cocktail (Roche, 11836145001) (*see* **Note 3**).
3. Dig-wash buffer: Wash buffer containing 0.05% digitonin (*see* **Note 4**).
4. Antibody buffer: Dig-wash buffer containing 0.05% digitonin, 2 mM EDTA, 0.1% BSA (*see* **Note 5**).
5. Dig-300 buffer: 20 mM HEPES pH 7.5, 300 mM NaCl, 0.5 mM spermidine, 0.01% digitonin, Complete Protease Inhibitor Cocktail (*see* **Note 6**).
6. Tagmentation buffer: Dig-300 buffer containing 10 mM $MgCl_2$ (*see* **Note 7**).
7. Digitonin (*see* **Note 8**).
8. Control mouse IgG (*see* **Note 9**).
9. Anti-mouse IgG (H&L) secondary antibody (ROCKLAND, 610-4120,) (*see* **Note 10**).
10. Proteinase K (20 mg/mL).
11. UltraPure Phenol-Chloroform-Isoamyl alcohol (25:24:1, v/v) (Invitrogen, 15593031).
12. Chloroform.
13. 100% ethanol.
14. NEBNext High-Fidelity 2× PCR Master mix (New England BioLabs, M0541S).
15. Agencourt AMPure XP beads (BECKMAN COULTER, A63882).
16. Tris-HCl (pH 8.0).
17. RNase A (25 μg/mL).
18. Qubit dsDNA HS assay kit (Invitrogen, Q32854).
19. High Sensitivity DNA Chips (Agilent Technologies, 5067-4626).
20. Sample-unique primer sets (*see* **Note 11**).

3 Methods

3.1 Preparation of Con A-Conjugated Dynabeads

1. To conjugate con A, 100 μL of streptavidin-conjugated Dynabeads MyOne T1 is washed with 1× PBS (pH 6.8) three times (*see* **Note 12**).
2. Resuspend the beads in 100 μL of 1× PBS (pH 6.8) containing 0.01% Tween-20.

3. Add 50 μL of 1× PBS (pH 6.8) containing biotin-conjugated con A (2.3 mg/mL) and 0.01% Tween-20 to the beads.
4. Rotate for 30 min at room temperature (RT).
5. Briefly spin down.
6. Remove the supernatant (*see* **Note 13**).
7. Resuspend the remaining beads in 100 μL of 1× PBS (pH 6.8) containing 0.01% Tween-20.
8. Store at 4 °C as con A-conjugated Dynabeads MyOne T1 (hereafter con A-MyOne T1).

3.2 pA-Tn5 Transposase Purification

1. Transform the expression vector, 3 × Flag-pA-Tn5-Fl, into *E. coli* strain BL21-pLysS Singles™ Competent Cells.
2. Culture in 50 mL of LB medium containing ampicillin (100 μg/mL) at 180 rpm at 37 °C until reaching OD600 = 0.9.
3. Add IPTG (final concentration at 0.25 mM).
4. Incubate at 180 rpm at 10 °C overnight.
5. Incubate at 180 rpm 23 °C for 4 h.
6. Briefly spin down the cells.
7. Remove the supernatant.
8. Resuspend the bacterial pellet in 20 mL of HEGX buffer containing Complete Protease Inhibitor Cocktail.
9. Lyse the cells by sonication on ice.
10. Centrifuge at 12,000 × g at 4 °C for 30 min.
11. Add 1/40 volume of 10% neutralized PEI.
12. Incubate at 4 °C for 30 min.
13. Filter the collected supernatant through a 0.45 μm mesh.
14. Add 7 mL of chitin slurry resin that is prewashed with HEGX buffer.
15. Incubated with rotation at 4 °C overnight.
16. Apply to an open chromatography column.
17. Wash with 20 mL of HEGX buffer six times.
18. Wash once with 14 mL of elution buffer.
19. Add an additional 7 mL of elution buffer.
20. Close the lid.
21. Rotate for 36–48 h at 4 °C.
22. Collect the elute.
23. Apply an additional 10.5 mL of elution buffer to the column.
24. Collect the elute (total ~17.5 mL).

25. Dialyze twice with 500 mL of dialysis buffer for more than 4 h at 4 °C.
26. Filter with a 0.45 μM filter.
27. Set a HiTrap SP1 gel filtration column in an ACTA start (set the flow rate at 0.5 mL/min).
28. Equilibrate the HiTrap SP1 gel filtration column with IEX-A buffer.
29. Load the sample to the column.
30. Wash the column with 5 mL of IEX-A buffer.
31. Elute with 50 mL 0.1–1.0 M NaCl gradient (collect 2 mL/fraction).
32. Check with SDS-PAGE and CBB staining.
33. Combine the fractions containing the eluted Tn5.
34. Dialyze with IEX-A buffer overnight at 4 °C (usually about 10 mL total).
35. Concentrate using Amicon Ultra-4 (30 K) to about 500 μL.
36. Adjust the concentration at 28.6 μM with IEX-A buffer (*see* **Note 14**).
37. Add an equal volume of 80% glycerol (pA-Tn5 concentration is 14.3 μM at this point).

3.3 pA-Tn5 Adapter Transposome Preparation

1. Mix mosaic end-adapter oligonucleotides (ME-A or ME-B) and mosaic end-reverse oligonucleotides (MErev) and NaCl as listed in Table 1 (*see* **Note 15**) (Fig. 1).
2. To make ME double-stranded (MEDS) oligonucleotides (100 μM), and anneal by heating at 95 °C for 5 min on a heat block followed by decreasing temperature at 0.1 °C/sec

Table 1
Amount of stock solutions to add to make Tn5MEDS-A or Tn5MEDS-B (100 mM)

Tn5MEDS-A (100 mM)	
Tn5ME-A (200 mM)	5 μL
Tn5MErev (200 mM)	5 μL
100 mM NaCl	0.5 μL
Tn5MEDS-B (100 mM)	
Tn5ME-B (200 mM)	5 μL
Tn5MErev (200 mM)	5 μL
100 mM NaCl	0.5 μL

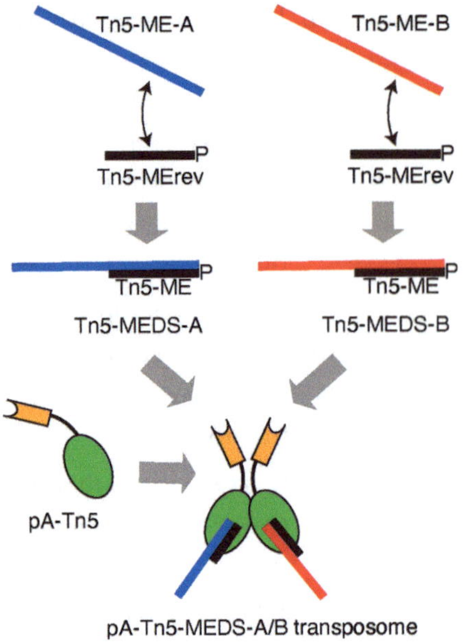

Fig. 1 Schematic image for steps in preparation of the pA-Tn5-MEDS-A/B transposome. Bars in blue or red represent oligonucleotides. See text for details of each procedure

(3% Ramp) to 75 °C and natural cooling down slowly to RT (<60 min) (*see* **Note 16**).

3. Mix equimolar of the pre-annealed Tn5MEDS-A and Tn5MEDS-B oligonucleotides (Tn5MEDS-A/B, 20 μL) (Fig. 1).
4. Mix 18 μL of Tn5MEDS-A/B with 126 μL of the pA-Tn5 fusion protein (14.3 μM) (Fig. 1).
5. Incubate the mixture with rotation for 1 h at RT.
6. Store at −20 °C as pA-Tn5-MEDS-A/B transposome (12.5 μM stock).

3.4 CUT&Tag

1. Add 10 μL (per ~100,000 cells of interest) of con A-MyOne T1 to 100 μL binding buffer in a 0.2 mL PCR tube.
2. Mix briefly and spin down.
3. Place the tubes on a magnet stand.
4. Remove the buffer.
5. Wash the beads with the binding buffer twice.
6. Resuspend in a 10 μL of the binding buffer.
7. Add 0.2 mL of the wash buffer.
8. Mix and briefly spin down.
9. Place the tubes on a magnet stand.

10. Remove the buffer.
11. Resuspend the beads in 50 μL of ice-cold antibody buffer.
12. Add 0.5 μg of primary antibody against the target protein or the same amount of control IgG (*see* **Note 17**).
13. Centrifuge the harvested ~100,000 cells of interest at 600 × g for 3 min at RT.
14. Aspirate the supernatant.
15. Resuspend the cells with the antibody buffer containing the beads and antibody from step 12.
16. Place the tube on an ELMI Intelli-mixer (PM-2 m, ELMI) (*see* **Note 18**).
17. Mix overnight at 4 °C (set with mode F8 at 30 rpm) (*see* **Notes 19** and **20**).
18. Briefly spin down.
19. Place the tubes on a magnet stand.
20. Remove the buffer.
21. Resuspend the beads in 100 μL of Dig-wash buffer (wash buffer containing 0.05% digitonin) containing the secondary antibody.
22. Place on an ELMI Intelli-mixer (ELMI) (*see* **Note 21**).
23. Mix for 60 min at RT.
24. Briefly spin down.
25. Place the tubes on a magnet stand (*see* **Note 22**).
26. Remove the buffer.
27. Wash the beads with 0.2 mL of Dig-wash buffer five times.
28. Resuspend the washed beads in 100 μL of Dig-300 buffer containing the pA-Tn5 mix (final concentration 50 nM) (1: 250 dilution of 12.5 μM pA-Tn5-MEDS-A/B transposome stock) (Fig. 2).
29. Mix gently.
30. Place the tubes on an ELMI Intelli-mixer (ELMI) (*see* **Note 23**).
31. Mix for 1 h at RT.
32. Briefly spin down.
33. Place the tubes on a magnet stand (*see* **Note 24**).
34. Remove the buffer.
35. Wash the beads with 0.2 mL of Dig-300 buffer twice.
36. Resuspend with 200 μL of tagmentation buffer.
37. Mix gently.

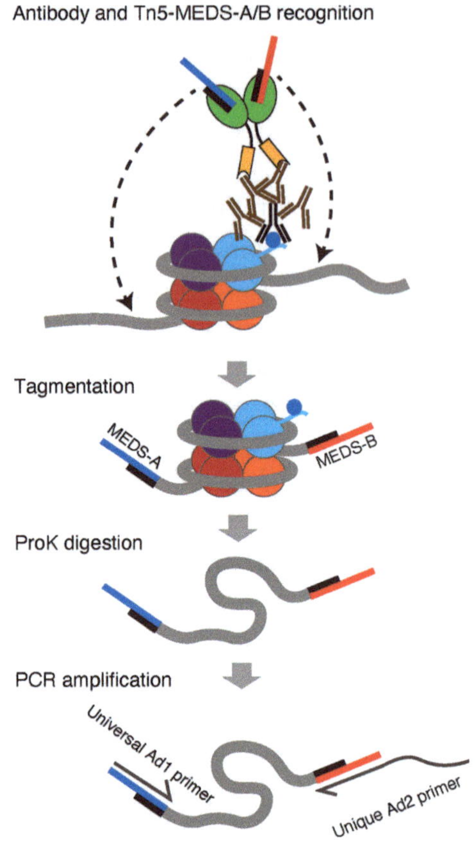

Fig. 2 Schematic image for steps in CUT&Tag. Bars in blue or red represent oligonucleotides, and bars in gray represent genomic DNA. Colored circles represent histones, and blue bubbles represent histone modification. See text for details of each procedure

38. Incubate for 1 h at 37 °C for adapter tagmentation reaction (Fig. 2).
39. Add 10 μL of 0.5 M EDTA, 3 μL of 10% SDS, and 2.5 μL of Proteinase K (20 mg/mL) to each sample to stop tagmentation and solubilize DNA fragments by de-crosslinking (Fig. 2).
40. Mix by vortexing.
41. Incubate at 37 °C overnight (*see* **Note 25**).
42. Transfer to a new 1.5 mL tube (*see* **Note 26**).
43. Add 300 μL of Phenol-Chloroform-Isoamyl alcohol (25: 24:1).
44. Mix by vortexing for a few seconds.
45. Centrifuge at 16,000 × g for 3 min at RT.
46. Transfer the aqueous upper layer to a new 1.5 mL tube (*see* **Note 27**).

47. Add 300 μL of chloroform.
48. Invert ~10 times.
49. Centrifuge at 16,000 × g for 3 min at RT.
50. Transfer the aqueous upper layer to a new 1.5 mL tube (*see* **Note 28**).
51. Add 750 μL of 100% ethanol.
52. Pipette well to mix.
53. Centrifuge at 16,000 × g for 15 min at 4 °C.
54. Remove the liquid by decantation carefully (*see* **Note 29**).
55. Rinse with 1 mL of 100% ethanol.
56. Centrifuge at 16,000 × g for 1 min at 4 °C.
57. Remove the liquid by decantation carefully.
58. Air-dry the purified DNA (*see* **Note 30**).
59. Dissolve in 30 μL of 10 mM Tris-HCl pH 8.0 containing 1 mM EDTA and 25 μg/mL RNase A (*see* **Note 31**).

3.5 PCR Amplification

1. Mix the adapter sequence-tagmented DNA fragments by PCR using 21 μL DNA (from 3.4 step 60), 2 μL of Universal Ad1 primer (10 μM), 2 μL uniquely barcoded Ad2 primers (index i7, 10 μM), and 25 μL NEBNext High-Fidelity 2× PCR Master mix (*see* **Note 32**).
2. The PCR cycle: 72 °C for 5 min, 98 °C for 30 s, 98 °C for 10 s, 63 °C for 10 s, and repeated 13 times, followed by 72 °C for 1 min.
3. To purify the PCR products, add 55 μL of Ampure XP beads (1.1 volume of the PCR product) (*see* **Note 33**).
4. Mix by pipetting.
5. Incubate for 10 min at RT.
6. Briefly spin down.
7. Place the tubes on a magnet stand until the buffer turns clear.
8. Remove the liquid.
9. Wash the beads carefully with 200 μL of 80% ethanol twice.
10. Briefly spin down.
11. Place the tubes on a magnet stand until the liquid turns clear.
12. Remove the liquid.
13. Air-dry the beads for 5–10 min.
14. Add 25 μL of 10 mM Tris-HCl (pH 8.0).
15. Vortex well.
16. Incubate for 5 min.
17. Briefly spin down.

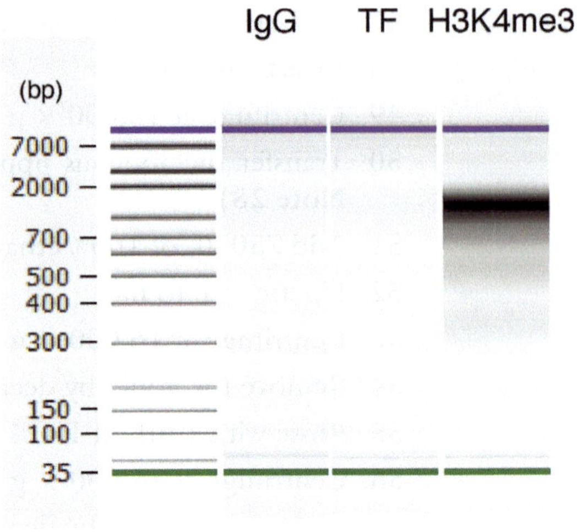

Fig. 3 Typical image of capillary electrophoresis for IgG, transcription factor (TF), and H3K4me3 using a bioanalyzer instrument

18. Place the tubes on a magnet stand until the liquid turns clear.
19. Transfer the liquid to a new tube.
20. Check the DNA concentration by using a Qubit 4 Fluorometer and Qubit dsDNA HS assay kit (*see* **Note 34**).
21. Check size distribution of the libraries by capillary electrophoresis using an Agilent 2100 Bioanalyzer Instrument and High Sensitivity DNA Chips (*see* **Note 35**) (Fig. 3).
22. Quantify the barcoded libraries by KAPA Library Quantification Kit, Complete kit for ABI Prism, Illumina platform, and sample-unique primer sets (*see* **Note 36**).
23. Perform paired-end or single-end sequencing of the barcoded libraries using an Illumina NGS sequencer.

4 Notes

1. Con A forms homotetramer at pH 5.6–7.0 and binds to receptors on the cell surface. This feature enables the magnetic beads to hold cells. However, it aggregates at pH 7.0 or above. Therefore, the pH of the PBS needs to be below 7.0. In this protocol, the pH is adjusted at 6.8.
2. Store the buffer at 4 °C for no longer than 6 months.
3. Store the buffer at 4 °C for no longer than 1 week.
4. Store the buffer at 4 °C for no longer than 2 days.
5. Store the buffer at 4 °C for no longer than 2 days.

6. The buffer can be stored at 4 °C for 1 week without adding digitonin. Digitonin should be added on the same day of the experiment.
7. The buffer should be used on the same day of the experiment.
8. Powder of digitonin is added to water and dissolved by heating at 98 °C for a couple of minutes. Digitonin solution can be stored at – 30 °C.
9. Use appropriate control IgG of the same host animal for the antibody against target antigens.
10. Use appropriate secondary antibodies for the primary antibody.
11. Sequences of the sample-specific primer sets are referred from the previous report [5].
12. Make sure the pH of PBS is adjusted at 6.8 (see Note 1).
13. A 100 μL of MyOne T1 slurry contains about 1 mg of beads, and 1 mg of T1 beads is capable of binding to 1100–1700 pmol of biotin. A 100 μL of recombinant solution contains a total of 230 μg con A recombinant protein (2.3 mg/mL), and each con A protein is conjugated with 4–8 mol biotin. Thus, 34.72–69.44 nmol biotin molecules are in the con A solution (2.3 mg/mL), which exceeds the binding capacity of 1 mg of MyOne T1 beads. The remaining supernatant con A solution can be quickly frozen with liquid nitrogen and stored at −80 °C for further use. Alternatively, a lower concentration of the con A solution may be used.
14. 28.6 μM; A280 = 2.69; 2.14 mg/mL.
15. Preparation of pA-Tn5-MEDS-A/B transposome (Fig. 1) was referred to and modified from the previously reported method [6].
16. A thermal cycler can be used for annealing oligonucleotides.
17. Use antibodies that can be used for immunostaining or ChIP-grade antibodies. Antibody for CUT&Tag is used at ten times higher concentration than for immunostaining.
18. A tube rotator can also be used.
19. Incubation can be at RT for 2 h or at 4 °C overnight.
20. Make sure beads are fully resuspended and all of the solutions are on the bottom of the tube.
21. A tube rotator can also be used.
22. Make sure to wait until the solution turns clear.
23. A tube rotator can also be used.
24. Make sure to wait until the solution turns clear.
25. A chunk of aggregated beads is usually formed at this step.

26. Use tubes with a tight lock cap or phase-lock tubes to avoid leakage of Phenol-Chloroform-Isoamyl alcohol.
27. Avoid contamination of protein layer, a thin middle layer between the aqueous and organic solvent layer. Use a tube with a tight lock cap to avoid leakage of alcohol.
28. Use a tube with a tight lock cap to avoid leakage of alcohol.
29. A visible DNA pellet is usually not observed.
30. Dry the ethanol completely. DNA pellet usually is not observed.
31. If observed DNA is not used immediately, store at – 30 °C for future use.
32. Sequences of the sample-specific primer sets are previously reported [5].
33. Before use, leave the bottle of Ampure XP beads at RT for longer than 30 min.
34. Concentrations of PCR products for histone modifications that are abundant in chromatin such as H3K4me3 are usually 1–20 ng/μL, while those for IgG control and TF samples are generally 0.1–2.0 ng/μL. If the concentration of the PCR product is below 0.1 ng/μL or detection limit, omit the sample and prepare a new sample.
35. Histone modifications abundant in chromatin such as H3K4me3 show DNA ladders, while IgG control and transcription factor (TF) samples usually do not. Figure 3 shows a typical DNA ladder pattern by a bioanalyzer when 1 × 105 cells are used. Ignore the higher-range PCR products over ~300 bp, which will not be sequenced by an NGS sequencer anyway. Concentrations of low-range PCR products (100 ~ 200 bp) are to be measured by qPCR.
36. If a pooled sample is to be sequenced, use the concentrations observed by qPCR.

References

1. Skene PJ, Henikoff S (2017) An efficient targeted nuclease strategy for high-resolution mapping of DNA binding sites, Elife:6. https://doi.org/10.7554/eLife.21856
2. Meers MP, Bryson TD, Henikoff JG, Henikoff S (2019) Improved CUT&RUN chromatin profiling tools. Elife:8. https://doi.org/10.7554/eLife.46314
3. Kaya-Okur HS, Wu SJ, Codomo CA, Pledger ES, Bryson TD, Henikoff JG, Ahmad K, Henikoff S (2019) CUT&Tag for efficient epigenomic profiling of small samples and single cells. Nat Commun 10(1):1930. https://doi.org/10.1038/s41467-019-09982-5
4. Kaya-Okur HS, Janssens DH, Henikoff JG, Ahmad K, Henikoff S (2020) Efficient low-cost

chromatin profiling with CUT&Tag. Nat Protoc 15(10):3264–3283. https://doi.org/10.1038/s41596-020-0373-x

5. Buenrostro JD, Wu B, Litzenburger UM, Ruff D, Gonzales ML, Snyder MP, Chang HY, Greenleaf WJ (2015) Single-cell chromatin accessibility reveals principles of regulatory variation. Nature 523(7561):486–490. https://doi.org/10.1038/nature14590

6. Picelli S, Björklund AK, Reinius B, Sagasser S, Winberg G, Sandberg R (2014) Tn5 transposase and tagmentation procedures for massively scaled sequencing projects. Genome Res 24(12):2033–2040. https://doi.org/10.1101/gr.177881.114

Chapter 17

Live Cell Synthetic Histone Acetylation by Chemical Catalyst

Shigehiro A. Kawashima and Motomu Kanai

Abstract

Posttranslational modifications (PTMs) of histones, such as lysine acetylation and ubiquitination, regulate chromatin structure and gene expression. In living organisms, histone PTMs are catalyzed by histone-modifying enzymes. Here, we describe an entirely chemical method to introduce histone modifications in living cells without genetic manipulation. The chemical catalyst PEG-LANA-DSSMe activates a thioester acetyl donor, N,S-diacetylcysteamine (NAC-Ac), and promotes regioselective, synthetic histone acetylation at H2BK120 in living cells.

Key words Catalyst, Histone, Acetylation

1 Introduction

Chemical modifications of histones, such as lysine acetylation and ubiquitination, play pivotal roles in epigenetic regulation of gene expression. Abnormal histone PTMs are tightly linked to numerous disorders in humans, such as type 2 diabetes, Alzheimer's disease, and various types of cancer [1–3]. Thus, chemical methods to install histone PTMs in live cells promise extraordinary potential in chromatin biology and medicine. In living organisms, the levels of histone PTMs are regulated by the interplay of two groups of histone-modifying enzymes, writers which introduce PTMs and erasers which remove them. In recent years, new chemical biology approaches that introduce histone PTMs entirely through artificial means without reliance on histone-modifying enzymes have emerged [4]. These include protein semisynthesis using native chemical ligation [5, 6], chemistry-coupled posttranslational mutagenesis [7, 8], and synthetic histone acetylation by chemical catalysis [9–12]. Among these, the chemical catalysis approach is unique because it does not require any genetic manipulations and thus has

Eisuke Gotoh (ed.), *Chromosome Analysis: Methods and Protocols*, Methods in Molecular Biology, vol. 2519,
https://doi.org/10.1007/978-1-0716-2433-3_17,

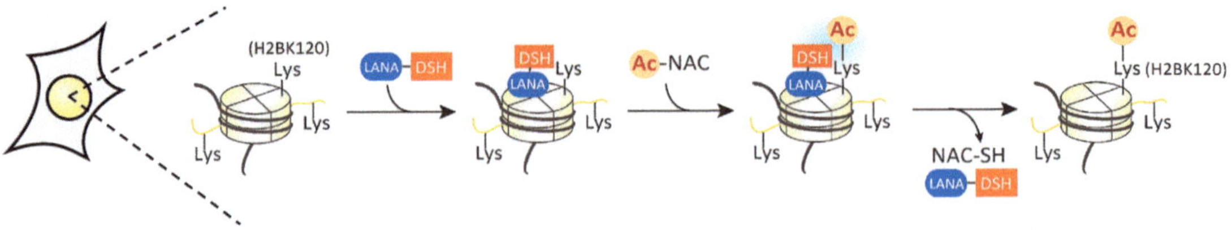

Fig. 1 Schematic of synthetic histone acetylation by chemical catalyst system. The catalyst, composed of a nucleosome-binding peptide (LANA) and a catalyst moiety (DSH), binds with nucleosome, activates a thioester acetyl donor (NAC-Ac), and promotes acetylation of H2BK120 in living cells

great potential for therapeutic use and broad epigenetics studies. Our newly synthesized chemical catalyst, PEG-LANA-DSSMe, binds with nucleosome's acidic patch and promotes regioselective, synthetic histone acetylation at H2BK120 in living cells [12] (Fig. 1). Importantly, the synthetic acetylation suppresses H2B ubiquitination, a critical epigenome mark in cell physiology. Here we describe how to promote synthetic histone acetylation reaction by chemical catalysts in living cells and how to analyze synthetic histone acetylation.

2 Materials

2.1 Synthetic Histone Acetylation Reaction in Living Cells

1. PEG-LANA-DSSMe (catalyst, Fig. 2a): 50 mM PEG-LANA-DSSMe in DMSO, stored in −30 °C freezer. Synthetic method was previously described [12] (*see* **Notes 1** and **2**).
2. NAC-Ac (acetyl donor, Fig. 2b): 5 M NAC-Ac in DMSO, stored in −30 °C freezer. Synthetic method was previously described [13] (*see* **Notes 1** and **3**).
3. Catalyst-donor solution: 500 μM PEG-LANA-DSSMe, 30 mM NAC-Ac, 2 mM tris(2-carboxyethyl) phosphine (TCEP), 0.5 mg/mL dextran-rhodamine (Thermo Fisher Scientific) in PBS.
4. Culture medium: Dulbecco's modified Eagle's medium (DMEM), 10% fetal bovine serum (FBS), 1× GlutaMax (Gibco), and antibiotics (100 U/mL streptomycin and 100 μg/mL penicillin).
5. Glass beads (≤106 μm; Sigma-Aldrich).
6. Six-well plate.
7. Nylon mesh-bottomed device: Make a container for sprinkling glass beads onto samples using a nylon mesh filter (~100 μm mesh size). Keep glass beads away from moisture; for example, store in a box containing silica gel [14].

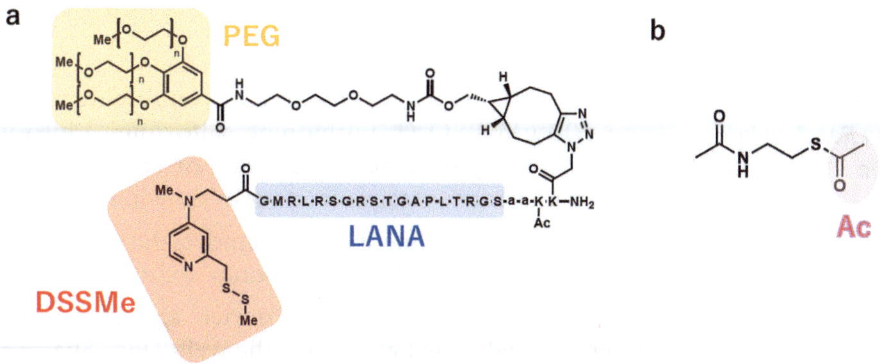

Fig. 2 Structure of a catalyst and an acetyl donor. (**a**) Structure of PEG-LANA-DSSMe. So far, optimal PEG length is $n = 13$. DSSMe is a prodrug of DSH and liberates the DSH catalyst under in-cell reducing conditions. (**b**) Structure of NAC-Ac, a partial structure of acetyl-CoA. The acetyl group (pink) is activated by the catalyst and transferred to a lysine residue on histones

2.2 Purification of Histones After Histone Acetylation Reaction

1. Hypotonic lysis buffer: 10 mM Tris-HCl (pH 8.0), 1 mM KCl, 1.5 mM $MgCl_2$, 1 mM DTT, 1 mM phenylmethylsulfonyl fluoride, protease inhibitor (e.g., cOmplete Mini, Roche).
2. Accutase (Innovative Cell Technologies).
3. 0.1% DAPI solution: 1 mg/mL DAPI solution (e.g., DOJINDO) is diluted into culture medium.
4. 35 μm filter: 5 mL polystyrene round-bottom tube with cell-strainer cap (FALCON).
5. Flow cytometer equipped with lasers for DAPI signals (e.g., 355 nm) and tetramethylrhodamine signals (e.g., 561 nm) (FACS Arial II, BD Biosciences).

2.3 Quantification of Acetylation Yield Using LC-MS/MS

1. Propionic anhydride solution: methanol/propionic anhydride, 3:1 [v/v] (*see* **Note 4**).
2. 1% ProteaseMAX solution: 10 mg/mL ProteaseMAX Surfactant (Promega) in 50 mM aqueous ammonium bicarbonate, stored in −30 °C freezer.
3. Trypsin gold solution: 1 mg/mL Trypsin Gold, Mass Spectrometry Grade (Promega) in 50 mM acetic acid, stored in −80 °C deep freezer.
4. Glu-C solution: 0.1 mg/mL Glu-C, Sequencing Grade (Promega) in distilled water, stored in −30 °C freezer.
5. Centrifugal evaporator (e.g., EYELA).
6. LC-MS/MS (e.g., TripleTOF 5600+ equipped with M5 MicroLC, SCIEX).

3 Methods

3.1 Synthetic Histone Acetylation Reaction in Living Cells

1. Plate 5 × 10^5 HeLa S3 cells in 2 mL culture medium containing 100 μM BSO (L-buthionine sulfoximine) into wells of a six-well plate (*see* **Note 5**). Use one or two wells in a six-well plate for efficient catalyst loading as shown in Fig. 3a.
2. After about 24 h incubation, remove the medium from the wells and add 100 μL catalyst-donor solution.
3. With a nylon mesh-bottomed device, sprinkle up to a monolayer of dried glass beads onto the wells (Fig. 3b).
4. Grab the six-well plate as shown in Fig. 3c and strike the plate against the hood tabletop ten times (Fig. 3d). Each time raise the dish ~20 cm above the tabletop (*see* **Note 6**).
5. Wash wells with 800 μL of culture medium several times until most beads are removed.
6. Add 800 μL of culture medium containing 30 mM NAC-Ac and 100 μM BSO.
7. Incubate cells for 2–8 h at 37 °C in CO_2 incubator.

3.2 Purification of Histones After Histone Acetylation Reaction

1. Wash cells by PBS twice.
2. Add 200 μL Accutase and incubate for 5 min at 37 °C.
3. Suspend the cells in 1 mL 0.1% DAPI solution, and transfer into low protein binding microcentrifuge tubes (e.g., Eppendorf).
4. Rotate the tubes for 15 min at room temperature.
5. Centrifuge the tubes (2000 rpm for 5 min at 4 °C) and remove supernatant.
6. Resuspend cells with 600 μL of culture medium.

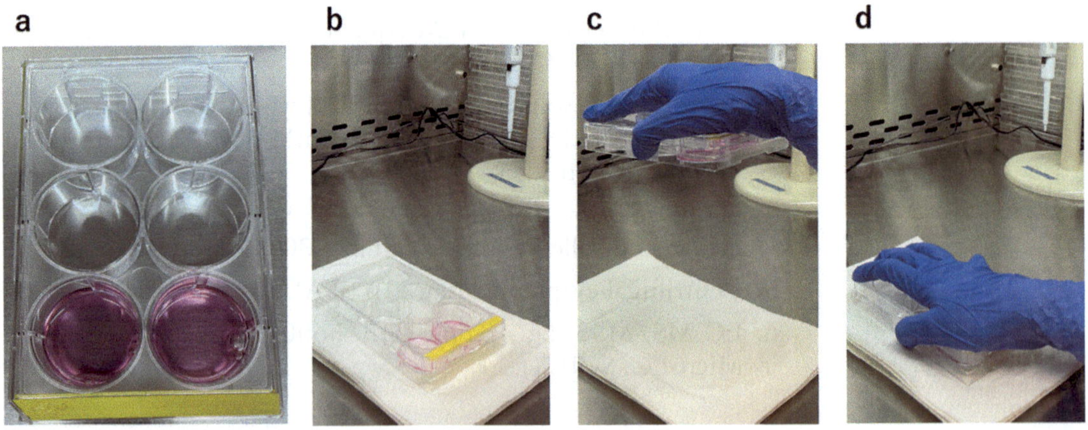

Fig. 3 Loading catalysts into live cells with glass beads. (**a**) Use one or two wells in a six-well plate. (**b**–**d**) After adding catalyst-donor solution and sprinkling dried glass beads, strike the six-well plate against the hood tabletop ten times

7. Prefilter cell solution through a 35 μm filter immediately before sorting.
8. Sort the living cells (DAPI-negative) by flow cytometer. If cells with high or low concentration of the catalyst are needed, sort cells with high or low tetramethylrhodamine signals.
9. Wash cells with PBS twice.
10. Resuspend cells with 1 mL hypotonic lysis buffer and keep the tubes on ice for 30 min.
11. Centrifuge the tubes (10,000 g for 10 min at 4 °C) and remove supernatant.
12. Resuspend pellet with 200 μL 0.4 N aqueous sulfuric acid and rotate the tubes at 4 °C for 30 min to overnight.
13. Centrifuge the tubes (16,000 g for 10 min at 4 °C) and transfer supernatant to new low protein binding microcentrifuge tubes.
14. Add 65 μL trichloroacetic acid and mix well by inverting tubes and keep the tubes on ice for 30 min to overnight.
15. Centrifuge the tubes (16,000 g for 10 min at 4 °C) and remove supernatant.
16. Wash the pellet by 300 μL ice-cold acetone twice.
17. Remove acetone completely by centrifugal evaporator (histone-containing pellet).
18. Analyze histone modifications by western blotting or LC-MS/MS analysis (*see* Subheading 3.3).

3.3 Quantification of Acetylation Yield Using LC-MS/MS

1. Add 10 μL distilled water to histone-containing pellet, tap the tubes, and incubate for 10 min at room temperature.
2. Add 10 μL 200 mM aqueous ammonium bicarbonate, 20 μL propionic anhydride solution, and 18 μL aqueous ammonium hydroxide (*see* **Note 7**), and incubate the tubes for 30 min at 25 °C.
3. Dry up the samples by centrifugal evaporator for >1 h at 80 °C.
4. Add 10 μL 0.1% ProteaseMAX solution in 50 mM aqueous ammonium bicarbonate and vortex well.
5. Add 90 μL 50 mM aqueous ammonium bicarbonate, 1 μL 1% ProteaseMAX solution, 1 μL Trypsin gold solution, and 10 μL Glu-C solution, and incubate the tubes for 3 h at 37 °C.
6. Add 25 μL 5% formic acid and vortex well to inactivate Trypsin or Glu-C.
7. Dry up by centrifugal evaporator for >1 h at 80 °C.
8. Add 15–20 μL 0.1% formic acid (*see* **Note 8**), centrifuge the tubes (14,000 rpm for 10 min at room temperature), and apply 5 μL to LC-MS/MS to analyze the stoichiometry of acetylated lysines of histones.

4 Notes

1. Synthesis of PEG-LANA-DSSMe and NAC-Ac needs expertise and equipment for organic chemistry and may be difficult in ordinary biology laboratories. For inquiries of these compounds, please contact the corresponding address.
2. The size of PEG can be changed. Note that conjugation of a longer PEG chain with the catalyst is advantageous for in-cell stability while disadvantageous for LANA-mediated nucleosome binding [12].
3. The type of acyl group can be changed. NAC-EtMal for lysine (ethyl)malonylation, NAC-N_3Ac for lysine (azido)acetylation, NAC-PA for lysine (propargyloxy)acetylation are now available [15].
4. Propionic anhydride solution must be prepared just before use.
5. Since the activity of the catalyst slightly decreased in the presence of high concentration of glutathione (GSH), a major reductant in cells, treatment of BSO (an inhibitor of γ-glutamylcysteine synthetase and depletes cellular GSH levels) increases acetylation efficiency by the catalyst system.
6. Between the efficiency of catalyst loading and cell survival rate is a trade-off. There is a need to optimize the tapping time/force and cell density, for individual cell types to achieve optimal catalyst loading and cell survival.
7. Confirm that pH of the solution is 7–10. If the solution is acidic, add more aqueous ammonium hydroxide.
8. When MS (mass spectrometry) intensity is too high, dilute the sample by 0.1% formic acid.

References

1. Hyun K, Jeon J, Park K, Kim J (2017) Writing, erasing and reading histone lysine methylations. Exp Mol Med 49(4):e324. https://doi.org/10.1038/emm.2017.11
2. Mirabella AC, Foster BM, Bartke T (2016) Chromatin deregulation in disease. Chromosoma 125(1):75–93. https://doi.org/10.1007/s00412-015-0530-0
3. Roadmap Epigenomics C, Kundaje A, Meuleman W, Ernst J, Bilenky M, Yen A, Heravi-Moussavi A, Kheradpour P, Zhang Z, Wang J, Ziller MJ, Amin V, Whitaker JW, Schultz MD, Ward LD, Sarkar A, Quon G, Sandstrom RS, Eaton ML, Wu YC, Pfenning AR, Wang X, Claussnitzer M, Liu Y, Coarfa C, Harris RA, Shoresh N, Epstein CB, Gjoneska E, Leung D, Xie W, Hawkins RD, Lister R, Hong C, Gascard P, Mungall AJ, Moore R, Chuah E, Tam A, Canfield TK, Hansen RS, Kaul R, Sabo PJ, Bansal MS, Carles A, Dixon JR, Farh KH, Feizi S, Karlic R, Kim AR, Kulkarni A, Li D, Lowdon R, Elliott G, Mercer TR, Neph SJ, Onuchic V, Polak P, Rajagopal N, Ray P, Sallari RC, Siebenthall KT, Sinnott-Armstrong NA, Stevens M, Thurman RE, Wu J, Zhang B, Zhou X, Beaudet AE, Boyer LA, De Jager PL, Farnham PJ, Fisher SJ, Haussler D, Jones SJ, Li W, Marra MA, McManus MT, Sunyaev S, Thomson JA, Tlsty TD, Tsai LH, Wang W, Waterland RA, Zhang MQ, Chadwick LH, Bernstein BE, Costello JF, Ecker JR, Hirst M, Meissner A, Milosavljevic A, Ren B, Stamatoyannopoulos JA, Wang T, Kellis M (2015) Integrative

analysis of 111 reference human epigenomes. Nature 518(7539):317–330. https://doi.org/10.1038/nature14248

4. Yamatsugu K, Kawashima SA, Kanai M (2018) Leading approaches in synthetic epigenetics for novel therapeutic strategies. Curr Opin Chem Biol 46:10–17. https://doi.org/10.1016/j.cbpa.2018.03.011
5. Muller MM, Muir TW (2015) Histones: at the crossroads of peptide and protein chemistry. Chem Rev 115(6):2296–2349. https://doi.org/10.1021/cr5003529
6. David Y, Vila-Perelló M, Verma S, Muir TW (2015) Chemical tagging and customizing of cellular chromatin states using ultrafast trans-splicing inteins. Nat Chem 7(5):394–402. https://doi.org/10.1038/nchem.2224
7. Wright TH, Bower BJ, Chalker JM, Bernardes GJL, Wiewiora R, Ng WL, Raj R, Faulkner S, Vallee MRJ, Phanumartwiwath A, Coleman OD, Thezenas ML, Khan M, Galan SRG, Lercher L, Schombs MW, Gerstberger S, Palm-Espling ME, Baldwin AJ, Kessler BM, Claridge TDW, Mohammed S, Davis BG (2016) Posttranslational mutagenesis: a chemical strategy for exploring protein side-chain diversity. Science 354(6312) doi:ARTN aag1465. https://doi.org/10.1126/science.aag1465
8. Yang A, Ha S, Ahn J, Kim R, Kim S, Lee Y, Kim J, Söll D, Lee HY, Park HS (2016) A chemical biology route to site-specific authentic protein modifications. Science 354(6312):623–626. https://doi.org/10.1126/science.aah4428
9. Ishiguro T, Amamoto Y, Tanabe K, Liu J, Kajino H, Fujimura A, Aoi Y, Osakabe A, Horikoshi N, Kurumizaka H, Yamatsugu K, Kawashima SA, Kanai M (2017) Synthetic chromatin acylation by an artificial catalyst system. Chem-Us 2(6):840–859. https://doi.org/10.1016/j.chempr.2017.04.002
10. Amamoto Y, Aoi Y, Nagashima N, Suto H, Yoshidome D, Arimura Y, Osakabe A, Kato D, Kurumizaka H, Kawashima SA, Yamatsugu K, Kanai M (2017) Synthetic posttranslational modifications: chemical catalyst-driven regioselective histone acylation of native chromatin. J Am Chem Soc 139(22):7568–7576. https://doi.org/10.1021/jacs.7b02138
11. Kajino H, Nagatani T, Oi M, Kujirai T, Kurumizaka H, Nishiyama A, Nakanishi M, Yamatsugu K, Kawashima SA, Kanai M (2020) Synthetic hyperacetylation of nucleosomal histones. RSC Chem Biol 1(2):56–59. https://doi.org/10.1039/D0CB00029A
12. Fujiwara Y, Yamanashi Y, Fujimura A, Sato Y, Kujirai T, Kurumizaka H, Kimura H, Yamatsugu K, Kawashima SA, Kanai M (2021) Live-cell epigenome manipulation by synthetic histone acetylation catalyst system. Proc Natl Acad Sci U S A 118(4). https://doi.org/10.1073/pnas.2019554118
13. Kulkarni RA, Worth AJ, Zengeya TT, Shrimp JH, Garlick JM, Roberts AM, Montgomery DC, Sourbier C, Gibbs BK, Mesaros C, Tsai YC, Das S, Chan KC, Zhou M, Andresson T, Weissman AM, Linehan WM, Blair IA, Snyder NW, Meier JL (2017) Discovering targets of non-enzymatic acylation by thioester reactivity profiling. Cell Chem Biol 24(2):231–242. https://doi.org/10.1016/j.chembiol.2017.01.002
14. Sato Y, Stasevich TJ, Kimura H (2018) Visualizing the dynamics of inactive X chromosomes in living cells using antibody-based fluorescent probes. Methods Mol Biol 1861:91–102. https://doi.org/10.1007/978-1-4939-8766-5_8
15. Hamajima W, Fujimura A, Fujiwara Y, Yamatsugu K, Kawashima SA, Kanai M (2019) Site-selective synthetic acylation of a target protein in living cells promoted by a chemical catalyst/donor system. ACS Chem Biol 14(6):1102–1109. https://doi.org/10.1021/acschembio.9b00102

Chapter 18

Histone Modification Analysis of Low-Mappability Regions

Naoko Yoshizawa-Sugata and Hisao Masai

Abstract

Posttranslational modifications of histone are intimately related to chromatin/chromosome-mediated cellular events. Among all, the roles of histone modifications including acetylation, methylation, ubiquitination, and SUMOylation of lysine or arginine residue of nucleosome core histones in gene expression have been intensively studied. Genome-wide profiles of histone modification marks revealed their combinatorial organization in the functional features of chromatin. Analysis of histone modification by chromatin immunoprecipitation (ChIP) is one of the standard assays to examine chromatin states. Although high-throughput sequencing analysis (ChIP-seq) is now widely conducted, classical ChIP-qPCR analysis has advantages in investigation of multiple histone modification marks at a target site simply through the use of relatively small numbers of cells. Since ChIP-qPCR is devoid of biases caused by overamplification and inaccurate mapping of sequencing reads, it is a more reliable quantification method than genome-wide ChIP-seq especially for analyses of the low-mappability regions, which harbor many repetitive sequences and/or highly homologous segmental multiplications as found in gene clusters. We have recently analyzed histone H3 and H4 modifications of the *Zscan4* family gene loci in an 880 kb gene cluster and found that the atypical enhancer-like structure is formed upon derepression of *Zscan4*. In this chapter, we describe the detailed protocols for histone modification ChIP-assay of repeat-enriched gene cluster regions. The protocol here we applied to mouse ES cells, but the protocol is perfectly applicable to human cultured cells and specimens.

Key words Chromatin immunoprecipitation (ChIP), Histone modification, Histone acetylation, Histone methylation, Gene expression, Gene cluster, Transposable elements, Low mappability, *Zscan4*, Rif1

1 Introduction

Histone modification is the one of the important epigenetic information and provides insights into chromatin states related to many key cellular events, such as gene expression, cell cycle regulation, DNA replication, and DNA damage responses and repair [1]. ChIP assay is used to analyze chromatin-bound proteins at target loci by using specific antibodies. Based on the accumulating knowledges of chromatin features of gene regulatory regions, histone modification analysis by using highly specific antibodies is frequently

Eisuke Gotoh (ed.), *Chromosome Analysis: Methods and Protocols*, Methods in Molecular Biology, vol. 2519,
https://doi.org/10.1007/978-1-0716-2433-3_18,

performed to detect the regulatory chromatin states [2]. The outline of the basic ChIP assay protocol for histone modification analysis is simple: shear the chromatin, enrich the modified histone-DNA complexes, purify DNA, and quantify it by quantitative PCR. In addition to the critical points in the protocol noted later in this chapter, the selection of antibodies and the design of the ChIP primers are important issues for histone modification analysis of gene cluster regions with low mappability where homologous segments, transposable elements, and/or low-complexity sequences exist. Therefore, we briefly describe the related information here.

Early analysis revealed that histone H3 and H4 acetylated at multiple residues are detected in active gene loci [3]. In addition to prominent lysin 27 acetylation (hereafter H3K27ac), lysine 9, lysine 14 (H3K9ac, H3K14ac) of H3 and lysine 8, and lysine 16 of H4 (H4K8ac and H4K16ac) are markedly detected at the promoter regions of actively transcribed genes. Histone H3 lysine 4 trimethylation (H3K4me3) also abundantly accumulates at promoters in yeast [4], and this feature is conserved in other higher eukaryotes. H3K4me1 modification is commonly detected at primed enhancer regions, and upon transcriptional activation, H3K27ac marks almost exclusively generated by CBP/P300 [5] are added to form active enhancers carrying H3K4me1/H3K27ac dual marks [6]. Gene repression is associated with two major markers, H3K9me3 and H3K27me3. H3K9me3 is a prerequisite for DNA methylation, a mark indicative of constitutive heterochromatin [7]. H3K27me3 is a silencing mark of the inactivated X chromosome in mammals and introduced by the Polycomb transcriptional repressor complex at promoter regions. Co-occupancy of histone marks in the same loci is commonly detected. For instance, in mouse ES cells, the regions carrying active H3K4me3 and silencing H3K27me3 marks are postulated as poised promoters. They are detected at the developmentally regulated genes which are activated at the later stages of cell differentiation [8], while such bivalent marks less frequently occur during early developmental stages in vivo [9]. Histone modification analysis by using multiple antibodies can illustrate the detailed chromatin states and uncover novel chromatin structures associated with specific combinations of histone modification that may be relevant to biological processes and diseases. Furthermore, histone modifications typically differ among the promoters, gene bodies, and transcription termination sites [10, 11], and multiple target sites should be analyzed to map precisely the chromatin states of the interest regions.

Recent progress in sequencing technology has allowed us to perform ChIP-sequencing analyses (ChIP-seq) to provide genome-wide views of chromatin states [12]. However, ChIP-seq is not suitable for quantitative analysis of some unusual genomic regions. Forty-six or 37.5% of the human or mouse whole genome, respectively, are composed of repetitive sequences, many of which are derived from transposable elements (TEs) and their degenerated remnants, mainly introduced by DNA transposons and retroviruses during evolution [13]. For example, the mouse genome contains at least 21 families of LTR TEs [14], and each TE family carries many copies of the sequences of high similarity; thus it makes difficult to locate TE-containing sequencing reads to unique positions in the reference genome. Gene clusters, composed of several to tens of numbers of family genes with highly homologous sequence, are often represented as low-mappability regions. In case that the sequence similarity is not restricted to gene bodies but also in their surrounding regions, sequencing reads cannot be efficiently aligned to the whole gene cluster (Fig. 1). There are ways to reduce the unmapped reads in sequence analysis process; selection and optimization of parameter settings of alignment tools increases mapping efficiency for TEs [15] (Fig. 1), and methods to quantify the TE-containing sequencing reads have been developed [16, 17]. Nevertheless, complete assignment of all reads to low-mappability regions is still difficult.

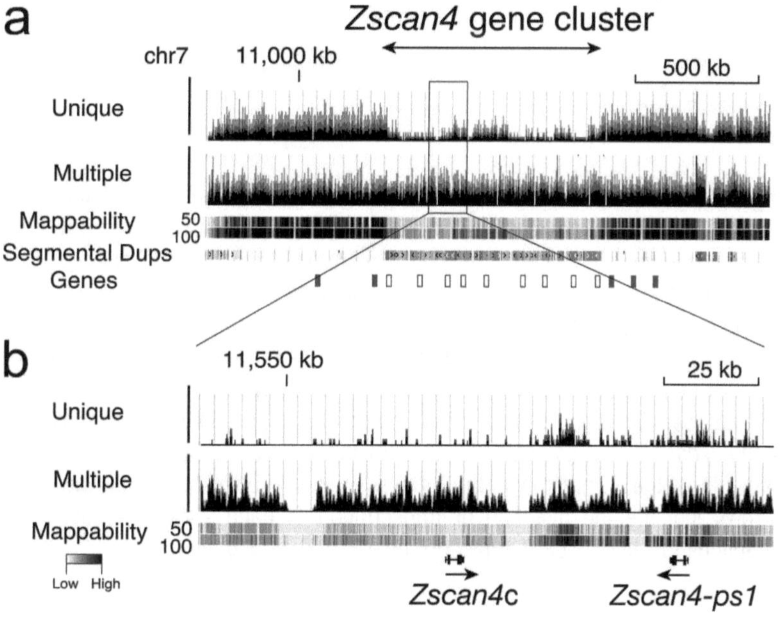

Fig. 1 Sequence reads aligned to a low-mappability region. Sequencing reads of ChIP-seq Input DNA were uniquely mapped (Unique) or mapped to multiple loci (Multiple) to mouse reference genome NCBI37/mm9 assembly by using Bowtie (ver. 0.12.7) or Bowtie2 (ver. 2.2.5) [23, 24]. The *Zscan4* gene cluster region is shown in the UCSC genome browser in A (2.25 Mb window) or B (150 kb window). Mappability of 50 or 100 mer sliding window is shown in lower tracks. In A, the positions of nine *Zscan4* family genes are indicated as open boxes, and nonfamily genes are as gray boxes. Note that whole *Zscan4* cluster is indicated to be highly probable for segmental duplication (in the lower track in A)

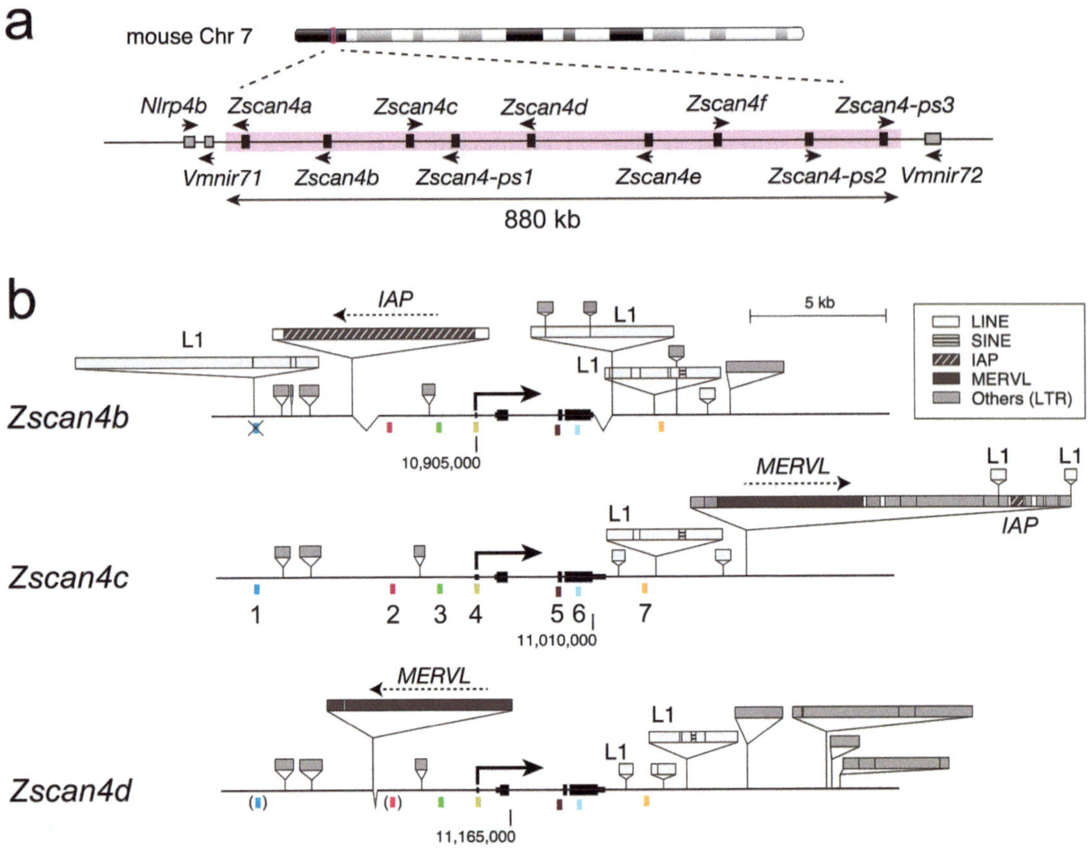

Fig. 2 Genomic structure of the *Zscan4* gene cluster. (**a**) The map of the nine family genes in the *Zscan4* gene cluster. The arrows indicate the direction of transcription. Gray boxes are three nonfamily genes adjacent to the gene cluster. (**b**) The genomic DNA structures of the *Zscan4* family gene loci. The *Zscan4b*, *Zscan4c*, and *Zscan4d* gene loci are shown as representatives. Insertions and deletions were identified from sequence homology between the prototype *Zscan4c* and other family members. Transposable elements and repeats (gray and shaded columns) were derived from RepeatMasker and RepeatMasker Viz. tracks of UCSC genome browser on mouse genome GRCm38/mm10 Assembly [25]. The color boxes indicate seven target sites of ChIP primers (*see* Fig. 3 and Table 2). The primers targeting 10 kb upstream sites (dark blue, 1) have no homology in the *Zscan4b* locus (indicated by a cross). In the *Zscan4d* locus, primers targeting 10 kb and 3.2 kb upstream sites (dark blue, 1, and magenta, 2; boxes in parentheses) have relatively lower homology (83–96% identity). Arrows indicate the direction of transcription of each *Zscan4* family gene. Dashed arrows indicate the directions of *IAP* or *MERVL* transcript near the *Zscan4* genes

In this chapter, we show a method for comprehensive ChIP-qPCR analysis of multiple histone modification in the *Zscan4* gene cluster region, as an example of low-mappability regions, by using either common or gene-specific 7 primer sets. *Zscan4* is required for telomere maintenance and specifically expressed in the two-cell stage of mouse zygote [18]. There are nine *Zscan4* family genes with high sequence similarity (96%<) and constitute the 880 kb gene cluster on the mouse chromosome 7 (Fig. 2a). *Zscan4* is expressed at the low level in mouse ES cells but transiently upregulated in rare populations (1 to several %), along with a group of two-cell specific transcripts and endogenous retrovirus *MERVL* [19]. Rif1, a DNA replication-timing regulator, is one of the suppressors of *Zscan4* expression in ES cells. Our recent ChIP assay shows that H3K27ac is significantly increased in broad regions, from at least 10 kb upstream to 1.9 kb downstream of the *Zscan4* gene in Rif1-depleted cells [20]. Unexpectedly, this enhancer-like structure lacks primed enhancer mark H3K4me1, whereas DNA methylation and H3K9me3 silencing marks are still maintained. None of the seven active histone H3 and H4 marks other than H3K27ac changed by Rif1 depletion in the *Zscan4* gene cluster. Three previous ChIP-seq studies resulted in contradictory observations. We have concluded that derepression of *Zscan4* caused by Rif1 depletion is associated with de novo formation of atypical broad enhancers with active H3K27ac mono-mark in semi-silencing mode.

2 Materials

Prepare all solutions with ultrapure water and reagents of analytical or molecular biology grade. Common equipment and materials are listed in the earlier subsection of the procedure.

2.1 Design of ChIP Primers

Software, programs, or web-based tools: To retrieve genomic sequences, use the UCSC Genome Browser (https://genome.ucsc.edu/), Ensembl (https://www.ensembl.org/), or others. To edit genomic sequences and conduct multiple alignment or homology search of family genes in gene clusters, commercial or public programs/tools are available. To design PCR primers, programs/tools such as Primer3 (https://primer3.ut.ee/) can be used.

2.2 Cell Culture and Preparation of Single Cell Suspension from Specimens

1. 10 cm plastic culture dishes.
2. Culture medium: Use appropriate medium for each cell line. For E14tg2a mouse ES cell, prepare DMEM supplemented with 15% fetal bovine serum validate for ES cells, 4 mM glutamine, 0.1 mM 2-mercaptoethanol, 1 mM sodium pyruvate, 0.1 mM non-essential amino acids, and 1000 U/mL of leukemia inhibitor factor.
3. Trypsin, collagenase, or other enzymes for cell detachment.
4. Reagents required for treatment of cells (*see* **Note 1**).
5. Homogenizer: (for human specimens) A manual or automatic homogenizer suitable for samples to use.

2.3 Stock Solutions

1. 10× phosphate-buffered saline (PBS): Dissolve 80 g of NaCl, 2 g of KCl, 28.9 g of $Na_2HPO_4 \cdot 12\ H_2O$, and 2 g of KH_2PO_4 with water. Make up the solution to 1 L and filtrate with a 0.22 μm filter.
2. 5 N NaOH: Take 40 mL of water and add 10.0 g of NaOH while stirring. Make up to 50 mL with water. Store in a plastic bottle.
3. 5 N KOH: Take 40 mL of water and add 14.03 g of KOH while stirring. Make up to 50 mL with water. Store in a plastic bottle.
4. 6 N HCl: Take 100 mL of water in a 300 mL glass beaker and slowly add 100 mL of concentrated HCl (12 N) while stirring gently with a magnetic stirrer.
5. 1 N HCl: Dilute 5 mL of 6 N HCl with 25 mL water.
6. 1 M HEPES-KOH (pH 7.5): Dissolve 23.83 g of HEPES with 80 mL water. Adjust pH to 7.5 with 5 N KOH. Make up the solution to 100 mL and filtrate with a 0.22 μm filter.
7. 5 M NaCl: Dissolve 58.44 g of NaCl with water and make up to 200 mL. Filtrate with a 0.22 μm filter.
8. 0.5 M EDTA (pH 8.0): Dissolve 37.22 g of ethylenediamine-N,N,N′,N′-tetraacetic acid disodium salt dihydrate (EDTA·2 H_2O) with 180 mL water and adjust pH to 8.0 with 5 N NaOH. Make up to 200 mL with water and autoclave.
9. 20% Triton X-100: Take 10 mL of Triton X-100 with syringe and make up to 50 mL with water.
10. 10% sodium deoxycholate: Weigh 1 g of sodium deoxycholate and make up to 10 mL with water.
11. 20% SDS: Dissolve 10 g of sodium dodecyl sulfate with 45 mL water and adjust pH to 7.2 with 1 N HCl and make up to 50 mL with water.
12. 1 M DTT: Dissolve 1.54 g of DL-dithiothreitol with water and make up to 10 mL. Aliquot and store at −30 °C.
13. 0.1 M Na_3VO_4: Dissolve 184 mg of sodium orthovanadate with water and make up to 10 mL. Aliquot and store at −30 °C.
14. 20% Tween-20: Take 10 mL Tween-20 with syringe and make up to 50 mL with water.
15. 20% Nonidet P-40: Take 10 mL of Nonidet P-40 with syringe and make up to 50 mL with water.
16. 4 M LiCl: Dissolve 8.478 g LiCl with water and make up to 50 mL. Filtrate with a 0.22 μm filter.
17. 1 M Tris-HCl (pH 8.0): Dissolve 60.57 g of tris (hydroxymethyl)aminomethane hydrochloride with 400 mL of water and adjust pH to 8.0 with 6 N HCl. Make up to 500 mL with water and autoclave.

18. 1 M Tris-HCl (pH 7.5): Dissolve 60.57 g of tris (hydroxymethyl)aminomethane hydrochloride with 400 mL of water and adjust pH to 7.5 with 6 N HCl. Make up to 500 mL with water and autoclave.
19. 5× Tris-borate buffer (TBE): Dissolve 54 g of tris (hydroxymethyl)aminomethane and 27.5 g of boric acid with water, and add 20 mL of 0.5 M EDTA. Make up to 1 L with water.
20. TE: Take 49.4 mL water in a 50 mL plastic tube and add 0.5 mL of 1 M Tris-HCl (pH 8.0) and 0.1 mL of 0.5 M EDTA (pH 8.0). Mix well and filtrate with a 0.22 μm filter.

2.4 ChIP (Day 1)

1. 1× PBS: Dilute 50 mL of 10× PBS with 400 mL of water. Adjust pH to 7.4 with 5 N NaOH and make up to 500 mL. Prepare just before use. Keep 200 mL at room temperature and 300 mL at 4 °C.
2. 1% formaldehyde: For up to seven 10 cm dishes, take 36 mL of PBS in a beaker and add 1 mL of concentrated formaldehyde (usually 36–38% w/vol), and mix well. Prepare immediately before use. Disposing the fluid wastes should follow institutional and national regulation.
3. 2 M glycine: Dissolve 7.5 g of glycine with water and fill up to 50 mL.
4. X-lysis buffer: Make with 50 mM HEPES-KOH (pH 7.5), 150 mM NaCl, 0.1% SDS, 0.1% sodium deoxycholate, and 1% Triton X-100 (for the readers' convenience, buffers for histone modification ChIP assay are summarized in Table 1, *see* Table 1).
5. X-lysis+ buffer: Prepare by adding 1 mM DTT, 1 mM Na_3VO_4, and protease inhibitors (e.g., 1 μg/ml pepstatin A, 1.5 μg/ml aprotinin, 1 μg/ml leupeptin, and 0.4 mM Pefabloc SC (Roche) at final concentration) to X-lysis buffer (*see* Table 1).
6. Nuclear lysis buffer: The same with X-lysis buffer except that 1% SDS, 1 mM DTT and protease inhibitors were added (*see* Table 1).
7. Nuclear lysis+ buffer: Prepare by adding 1 mM DTT, 1 mM Na_3VO_4, and protease inhibitors to nuclear lysis buffer (*see* Table 1).
8. 2× glycerol loading buffer: Take 0.5 mL of glycerol with wide-pore pipet tip or syringe to a 1.5 mL tube and add 0.5 mL of TE. Mix by rotation.
9. 1× TBE: Dilute 200 mL of 5× TBE buffer with 800 mL of water and mix well.

Table 1
Buffer preparation table for histone modification ChIP assay

Buffers and compositions	Stock buffer concentration	Stock buffer volume	Total volume
X-lysis buffer (0.1% SDS)			500 mL
50 mM HEPES-KOH pH 7.5	1 M	25 mL	
150 mM NaCl	5 M	15 mL	
1 mM EDTA	0.5 M	1 mL	
1% triton X-100	20%	25 mL	
0.1% sodium deoxycholate	10%	5 mL	
0.1% SDS	20%	2.5 mL	
Water		427 mL	
X-lysis+ buffer			50.8 mL
X-lysis buffer		50 mL	
1 mM DTT	1 M	50 μL	
1 mM Na_3VO_4	0.1 M	0.5 mL	
Protease inhibitors	1000×	50 μL each	
X-lysis+ buffer without DTT			50.7 mL
X-lysis buffer		50 mL	
1 mM Na_3VO_4	0.1 M	0.5 mL	
Protease inhibitors	1000×	50 μL each	
ChIP binding buffer			50.7 mL
X-lysis buffer		50 mL	
0.5 mM DTT	1 M	25 μL	
1 mM Na_3VO_4	0.1 M	0.5 mL	
Protease inhibitors	1000×	50 μL each	
Nuclear lysis buffer (1% SDS)			250 mL
50 mM HEPES-KOH pH 7.5	1 M	12.5 mL	
150 mM NaCl	5 M	7.5 mL	
1 mM EDTA	0.5 M	0.5 mL	
1% triton X-100	20%	12.5 mL	
0.1% sodium deoxycholate	10%	2.5 mL	
1% SDS	20%	12.5 mL	
Water		202 mL	

(continued)

Table 1
(continued)

Buffers and compositions	Stock buffer concentration	Stock buffer volume	Total volume
Nuclear lysis+ buffer			10.2 mL
Nuclear lysis buffer		10 mL	
1 mM DTT	1 M	10 μL	
1 mM Na_3VO_4	0.1 M	0.1 mL	
Protease inhibitors	1000×	10 μL each	
ChIP wash buffer 1 (0.35 M NaCl)			250 mL
50 mM HEPES-KOH pH 7.5	1 M	12.5 mL	
350 mM NaCl	5 M	17.5 mL	
1 mM EDTA	0.5 M	0.5 mL	
1% triton X-100	20%	12.5 mL	
0.1% sodium deoxycholate	10%	2.5 mL	
1% SDS	20%	1.25 mL	
Water		203.3 mL	
ChIP wash buffer 2 (0.25 M LiCl)			250 mL
10 mM Tris-HCl pH 8.0	1 M	2.5 mL	
250 mM LiCl	4 M	15.6 mL	
1 mM EDTA	0.5 M	0.5 mL	
0.5% Nonidet P-40	20%	6.25 mL	
0.5% sodium deoxycholate	10%	12.5 mL	
Water		213 mL	
ChIP elution buffer			250 mL
50 mM Tris-HCl pH 7.5	1 M	12.5 mL	
10 mM EDTA	0.5 M	5 mL	
1% SDS	20%	12.5 mL	
Water		220 mL	
TNE buffer			50 mL
20 mM Tris-HCl pH 8.0	1 M	1 mL	
100 mM NaCl	5 M	1 mL	
1 mM EDTA	0.5 M	0.1 mL	
Water		47.9 mL	

10. 1.5% agarose gel in TBE: Add 1.5 g of agarose of molecular biology grade to 100 mL of 1× TBE and dissolve by heating by microwave. Cool to 60 °C and transfer warmed gels to plastic container with combs [21].
11. X-lysis+ buffer minus DTT: same with X-lysis+ buffer except that DTT was omitted (*see* Table 1).
12. ChIP binding buffer: The same with X-lysis+ buffer except that 0.5 mM DTT was added (*see* Table 1).
13. PBS/0.02% Tween-20 (PBS-Tw): For five antibodies, take 50 mL of 1× PBS to a 50 mL plastic bottle and add 50 μL of 20% Tween-20. Mix well and filtrate with a 0.22 μm syringe filter.
14. PBS/0.1% Triton X-100 (PBS-Tx): For five antibodies, take 50 mL of 1× PBS to a 50 mL plastic bottle and add 250 μL of 20% Triton X-100. Mix well and filtrate with a 0.22 μm syringe filter.
15. Shakers: To shake culture dishes and 15 mL tubes, a shaker with a horizontally swirling plate or a see-saw plate moving in low speed is preferred. If not available, shake mildly by manual for 30 s every 5 min during the incubation periods.
16. Silicone rubber-attached scrapers (*see* **Note 2**) (Fig. 4).
17. DNA low-binding tubes and low-retention pipet tips (*see* **Note 3**).
18. Centrifuges: Use a refrigerated centrifuge machine with swinging rotors for 15 mL and 1.5 mL tubes or a refrigerated high-speed centrifuge with either a swinging rotor or an angle rotor.
19. Rotator for 1.5 mL and 15 mL tubes in a cold chamber.
20. Sonicator: Probe-type, water bath-type, or indirect acoustic focused-type sonicator.
21. Magnetic stands for 1.5 mL and 15 mL tubes.
22. Antibodies for specific histone modification.
23. Antibody-bound magnetic beads: Calculate required amounts of antibodies and magnetic beads beforehand. For pan histone or abundant modification, use 20 μL of magnetic beads (e.g., 0.6 mg of beads with binding capacity of 4.8 μg human IgG for Dynabeads® Protein G) and 1 μg of antibody per ChIP replicate (*see* **Note 4**). Wash beads (up to 300 μL) with 0.5 mL (up to 1 mL) of PBS-0.02% Tween (PBS-Tw) twice and conjugate with corresponding amounts of antibody against modified histone in 200 μL (up to 1 mL) of PBS-Tw in 1.5 mL low DNA-binding tubes (*see* **Note 5**). Incubate for 2–6 h, during **steps 3–8** in Subheading 3.3. Wash unbound antibodies out with 0.5 mL of PBS-0.1% Triton X-100 (PBS-Tx) three times and suspend homogeneously with ChIP binding buffer.

Equally distribute beads suspension into new 1.5 mL tubes in ChIP experiment number (sample number × replicates) and store on ice until use (*see* **Note 6**). Follow the manufacturer's instructions of magnetic beads.

2.5 ChIP (Day 2)

1. ChIP wash buffer 1 (NaCl): X-lysis buffer containing 350 mM NaCl (*see* Table 1).
2. ChIP wash buffer 2 (LiCl): 10 mM Tris-HCl pH 7.5, 0.25 M LiCl, 1 mM EDTA, 0.5% Nonidet P-40, and 0.5% sodium deoxycholate (*see* Table 1).
3. ChIP elution buffer: 50 mM Tris-HCl pH 7.5, 10 mM EDTA, and 1% SDS (*see* Table 1).
4. RNase A and proteinase K.
5. Phenol/chloroform/isoamyl alcohol: Mix tris-saturated phenol, chloroform, and isoamyl alcohol at 25:24:1 ratio [21].
6. TNE buffer: Prepare with 20 mM Tris-HCl (pH 8.0), 100 mM NaCl, and 1 mM EDTA (pH 8.0). Mix well and filtrate with a 0.22 μm filter (*see* Table 1).
7. 2-propanol.
8. 3 M sodium acetate: Dissolve 81.62 g of sodium acetate trihydrate in 150 mL of water and adjust to pH 5.2 by adding acetic acid. Fill up to 200 mL and filtrate with a 0.22 μm filter.
9. 20 mg/mL glycogen solution.
10. 75% ethanol: Take 37.5 mL of ethanol and make up to 50 mL with water.
11. DNA purification column kit: Commercially available DNA purification kit, composed of silica membrane-based mini columns and buffers.
12. Tris-HCl buffer (TB): Take 9.9 mL of water in a 15 mL plastic tube and add 0.1 mL of Tris-HCl (pH 8.0). Mix well and filtrate with a 0.22 μm filter.
13. Hybridization oven with shaker, or thermal mixer.

2.6 Quantitative Real-Time PCR (qPCR)

1. 10 μM primer set stock solution: Make 100 μM solution for each primer. Add each 5 μL of forward and reverse primer to 40 μL of TB and mix. Store at −20 °C.
2. Real-time PCR enzyme and reagents (*see* **Note 7**).
3. Real-time PCR system: The systems equipped with a charge-coupled device (CCD) camera and at least single detection filter for SYBR Green fluorescence or its derivatives.
4. Multi-well plate: Use 96-well or 384-well plates compatible with the real-time PCR system.

3 Methods

Carry out all procedures at room temperature unless otherwise specified.

3.1 Design of ChIP Primers

Design and validate qPCR primers for the regions of gene clusters harboring highly homologous family genes. For this, we recommend to retrieve sequence sets of all family genes in a gene cluster with upstream and downstream regions. Genomic structure of *Zscan4* family genes and the sites of primers were shown as an example (Fig. 2b; [20]).

3.1.1 Retrieving Genomic Sequences

If family genes are annotated in the reference genome, retrieve all family gene sequences separately, with extra upstream and downstream sequences (e.g., 10 kb) (*see* **Note 8**). In case the gene cluster has no complete gene annotation, find all family genes from the whole genomic sequences by BLAST (https://blast.ncbi.nlm.nih.gov/Blast.cgi). To download surrounding sequences of such gene clusters, consider their low sequence conservation and determine the extent of the segment to be retrieved.

3.1.2 Multiple Alignments, Homology Search and Selecting a Prototype of Family Genes

All family gene sets with neighboring sequences are subjected to multiple alignment. Pay attention to genes directions. Identify the conserved regions and clarify deletion or insertion of unrelated sequences in specific family members. According to clustering analysis result (by CLASTALW (https://www.genome.jp/tools-bin/clustalw or others), illustrate overall sequence similarities among family members and choose a well-conserved, full-length family gene as a prototype (*see* **Note 9**).

3.1.3 Selection of Primer Target Sites

1. All primers: Follow the general rules for designing primers (*see* **Note 10**). Basically, avoid to set primers in repeat regions, although the target amplicon sequences between two primers can include short repeats.
2. Common primers: Choose pairs of candidate primers in conserved sequences among family gene loci for the regions of interest. Primer sequence can include a short stretch of repeats at the part of 5′ end if it is well conserved among family genes. Search sequence homology of primers within all family gene loci. Check amplicon size for each family member as well, to confirm if long TEs or repeats are not inserted between primer pairs (*see* **Note 11**). An example of homology search results including primer target sites and amplicon sizes of all family genes in the *Zscan4* gene cluster is shown in Table 2.
3. Unique family gene-specific primers: If required, select the exclusive sites which is only identified in one or a small fraction

Table 2
ChIP primer target regions and probability of PCR amplification of each family gene locus in the *Zscan4* gene cluster

	***Zscan4* family genes**									
Target site	**Amplicons**	**a**	**b**	**c**	**d**	**e**	**f**	**ps1**	**ps2**	**ps3**
10 kb upstream (1)	Location of primer F	–	−7330	−10160	(−26059)		−10137	−12748	(−10099)	−15725
	Location of primer R	(−9650)	–	−9935	(−25837)	(−6968)	−9912	–	−9873	−15500
	Length (bp)	–	–	226	223	–	226	–	227	226
	Probability	–	–	+++	++	–	+++	–	++	+++
3.2 kb upstream (2)	Location of primer F	(−3494)	−3261	−3300	(−2775)	−3249	−3277	(−3365)	−3250	−8808
	Location of primer R	–	−3076	−3116	−2590	−3064	−3093	–	−3066	−8624
	Length (bp)	–	185	184	185	185	184	–	184	184
	Probability	–	+++	+++	++	+++	+++		+++	+++
1.4 kb upstream (3)	Location of primer F	−1466	−1439	−1455	−1146	−1438	−1431	−1486	−1430	−1463
	Location of primer R	−1277	−1249	−1274	−953	−1250	−1251	−1298	−1250	−1282
	Length (bp)	189	190	181	193	188	180	188	180	181
	Probability	+++	+++	+++	+++	+++	+++	+++	+++	+++
TSS (4)	Location of primer F	−144	−130	−153	−164	−130	−130	−164	−130	−153
	Location of primer R	125	140	117	106	140	140	106	140	117
	Length (bp)	269	270	270	270	270	270	270	270	270
	Probability	+++	+++	+++	+++	+++	+++	+++	+++	+++

The location and amplicon size of ChIP target sites in the nine *Zscan4* family gene loci in correspondence to the prototype gene *Zscan4c*. Four out of seven common primer sets in the upstream regions ("1–4" in Figs. 2 and 3) are shown. "Location of primer F (or R)" indicate the positions where the 5' ends of forward (F) or reverse (R) primers match with >90% sequence identity (0 represents the transcription start site of the gene). Numbers in parenthesis indicate the locations with 80–90% identity to primers, while those in double parenthesis indicate the locations with 70–80% identity to primers. –, less than 70% identity or no homology. The expected PCR amplification efficiency of the target regions was estimated by sequence homology to primers and target amplicon size, which was indicated as "Probability" of family gene loci as gene-specific primers (*see* **Note 12**). The border sites of conserved sequences and gene-specific TEs/repeats can be selected as candidates.

3.1.4 Validation of ChIP Primers

Validate ChIP primers by real-time PCR using the sonicated chromatin (ChIP "Input" DNA) as a template. Following the instrument manuals, check that melting peak of each primer set is single and without major shoulders and not multimodal. Peaks are often broader compared to general PCR primers due to variation of nucleotide composition and sequence length of the amplicons derived from multiple family genes, and that is usually not a problem (*see* **Note 13**).

3.2 Cell Culture and Preparation of Single Cell Suspension from Specimens

Culture cells at subconfluent density and change medium at appropriate timing throughout the experiments. Treat cells with reagents for gene depletion or stimuli if required. Obtain adequate cell numbers (*see* **Note 14**). For example, prepare two 10 cm culture plate of E14tg2a mouse ES cells at semiconfluent density for three ChIP assays of abundant histone modifications in three ChIP replicates. For human specimens, dissociate tissues or aggregated samples into single-cell suspension by combining mechanical dissociation and enzymatic treatment [22]. Process human cells similarly as in following sections, except for fixing, quenching, and washing cells in suspension (in Subheading 3.3, **steps 1–3**). Use human clinical samples under the national ethical regulation on research on human subjects and materials. Treat potentially infectious clinical samples carefully in biosafety level 2 (BSL2) containment facilities, at least until cells are fixed with formaldehyde.

3.3 ChIP (Day 1)

1. Fixation: Following protocol is for four 10 cm plates. After washing with 5 mL of PBS per dish two times, gently add 5 mL of 1% formaldehyde per dish and incubate at r.t. for 15 min with gentle swirling or shaking on shaker (40 rpm for 2.5 cm stroke).
2. Quenching crosslinkers: Add 0.35 mL of 2 M glycine per dish and shake for 10 min.
3. Washing cells: Remove solution and wash cells with 5 mL of cold PBS two times. Add 1 mL of cold PBS and collect cells by a scraper to one end of a dish. Harvest cell suspensions with 1 mL low-retention pipet tip into 15 ml or 50 mL polypropylene tube (*see* **Note 2**). Wash a plate with 3 mL of cold PBS and combine remaining cells to a tube. Centrifuge cells at 440 × g for 10 min at 4 °C and remove supernatants. Wash cells by gently pipetting with 5 mL of cold PBS, centrifuge, and remove supernatants.
4. Nuclei isolation: Resuspend cell pellets with 3 mL of X-lysis+ buffer and incubate at 4 °C for 15 min with gentle shaking

(e.g., 50 rpm with 4° tilt angle or 1 cm of stroke). Centrifuge at 440 × g for 10 min at 4 °C and remove supernatants. Repeat this step twice.

5. Nuclear lysis: Resuspend with 3 mL of nuclear lysis buffer+ and incubate at 4 °C for 15 min with gentle shaking as in step 3. Centrifuge at 440 × g for 10 min at 4 °C (*see* **Note 15**) and remove supernatants.
6. Washing chromatin: Resuspend with 3 mL of X-lysis+ buffer and incubate at 4 °C for 15 min with gentle shaking. Centrifuge at 440 × g for 10 min at 4 °C (*see* **Note 15**) and remove supernatants. Repeat this step (total two times). Pellets can be stored at −80 °C for short period.
7. Sonication: Resuspend chromatin pellets with 0.9 mL of X-lysis + buffer in 15 mL polypropylene tubes, and shear by a sonicator in optimized condition (*see* **Note 16**). We usually conduct combination protocol (sequential processing by using two sonicators), and the following are one of the optimized conditions: (1) sonicate samples by Sonifier 250D with tapered 1/8-inch-diameter microtip (Branson) on ice at 20% intensity with 50% frequency (1 s on and 1 s off) for 1 min, for 2–3 times with 2-min intervals. (2) Transfer the sonicates to 1 mL glass tube and fill up to 1 mL with X-lysis+ buffer. (3) Process by focused-ultrasonicator S-220 (Covaris) at power 150 W, Duty Factor 20, for 12–15 min at 4 °C. Confirm the size of sheared chromatin by agarose gel electrophoresis. The size must be ranging from 100 to 400 bp with average size at 200 bp (*see* **Note 17**).
8. Preclear: Dilute the sonicates with 1 mL of X-lysis+ buffer minus DTT and add ChIP binding buffer (final 0.5 mM DTT) to adjust the concentration at approximate 1×10^6 cells per 0.5 ml. Centrifuge at 39,000 × g for 10 min. Transfer the supernatants to 15 mL tube and add 50 μL of PBS-Tween washed magnetic beads per 1×10^7 cells and incubate for 1–6 h at 4 °C on a rotator. Preclear chromatin while antibodies are bound to magnetic beads (Subheading 2.3, **step 23**).
9. ChIP: Centrifuge sheared chromatin with beads at 1750 × g for 10 min in a swing rotor and put the tube on a magnetic stand. Collect the supernatants to a new tube and leave preclear beads and the precipitates (if appeared). Take one hundredth of sonicates as Input DNA and store at −80 °C. Immediately before starting immunoprecipitation, remove PBS-Tx buffer of antibody-conjugated beads (*see* Subheading 2.3, **step 23**) on magnetic stand and transfer tubes on ice. Distribute 0.5 mL of precleared chromatin to antibody-bound beads and incubate at 4 °C overnight (12–16 h) on a rotator.

3.4 ChIP (Day 2)

1. ChIP Wash 1: Remove unbound fractions (supernatants) on magnetic stand. Add 0.5 mL of X-lysis buffer and rotate at 4 °C for 5 min. Remove supernatants on magnetic stand. Repeat this step twice (three times in total). Following wash steps are conducted in the same way.
2. ChIP Wash 2: Wash beads with 0.5 mL of ChIP wash buffer 1 (0.35 M NaCl) once.
3. ChIP Wash 3: Wash beads with 0.5 mL of ChIP wash buffer 2 (0.25 M LiCl) once.
4. ChIP Wash 4: Wash beads with 0.5 mL of TE. Centrifuge the tube at 400 × g at 4 °C for 30 s and set on a magnetic stand and remove supernatants completely.
5. Elution: Add 0.3 mL of ChIP elution buffer to beads and incubate at 37 °C for 30 min on rotary mixer with constant tapping or on Thermomixer at 500 rpm. Transfer the ChIP eluates to new 1.5 mL tubes.
6. De-crosslinking: Incubate the eluates and Input DNA (*see* Subheading 3.3, **step 9**) at 65 °C for 6 h in a hybridization oven with constant shaking at 60 rpm. Process both samples similarly in the following steps.
7. RNase and ProK treatment: Add RNaseA (50 μg/ mL) and incubate at 37 °C for 1 h. Place at room temperature, then add proteinase K (0.5 mg/mL) and incubate at 50 °C for 2 h.
8. DNA extraction: Add the same volume of phenol/chloroform/isoamyl alcohol, mix vigorously, and recover the supernatants. Re-extract the organic phase with 1× volume of TNE buffer. Combine aqueous phase and add 20 μg of glycogen, 0.3 M sodium acetate, and 1× volume of 2-propanol per ChIP. Mix well and incubate at −80 °C for at least 30 min. The sample can be stored for later analysis (*see* **Note 18**).
9. DNA purification: Centrifuge at 17,000 × g for 30 min. Wash pellets with 1 mL of 75% ethanol and repeat (two times in total). Air-dry DNA pellets and resuspend with 50 μL of TE. R-epurify DNA by silica membrane-based mini column and elute with 100 μL of TB. Use the sample as soon as possible, although they can be stored at −20 °C for a short period (less than 5 days) (*see* **Note 19**).

3.5 Real-Time PCR

1. Standard curve: Dilute Input DNA with TB by tenfold dilution series for three times. To decide dilution range, make standard curves by using dilution series for each primer set in Subheading 3.1 (*see* **Note 20**).
2. ChIP DNA dilution (if required): In the same PCR reactions in 1, check all purified ChIP samples are in the range of standard curves. No replicate is needed for these check. Decide dilution factors, of ChIP DNA, or use original solutions.

3. Quantification of ChIP DNA: Follow the instructions of real-time PCR reagents for preparation of premix. For example, mix 0.2 μL of primer pair stock solution and 5 μL of 2× enzyme-reagent mix, and fill up with water to 9 μL per reaction. Make premix for each primer set and dispense 9 μL per well in 96-well plate (*see* **Note 21**). Add 1 μL of original or diluted ChIP DNA to each well. After sealing with a plate film and spin down, set a plate on real-time PCR system.
4. Calculation of ChIP efficiency: Calculate the ratio of ChIP DNA to Input DNA (*see* **Note 22**). The values quantified by Input DNA in step 3 are compensated by factors (ratio of used DNA for PCR reaction in total DNA). In general, analyze ChIP efficiency in technical triplicates in multiple independent experiments, or in biological triplicates, and show the average values with standard deviations. The results of four histone H3 modifications and two H4 modifications in the *Zscan4* gene cluster region by using common primers (targeting at least 5 out of 9 family genes) are shown in Fig. 3.

4 Notes

1. To assess the effects of gene knockdown by RNAi or knockout by Cre recombinase-fused estrogen receptor (Cre-ER)/LoxP system, treat cells with siRNA/shRNA or 4-hydroxytamoxifen, respectively. Similarly, treat cells with auxin or other reagents for proteasome-mediated protein degradation system, if required. If the treatments inhibit cell growth, increase the culture scale to harvest the same cell numbers for each group.
2. After fixation and dispersion, cells tend to float in buffers and attach to the plastic surface, probably because the cell surface becomes hydrophobic. To minimize the loss, collecting cells in small volumes as clots by using a silicone rubber scraper is recommended. Scrapers can be obtained commercially or assembled in advance by attaching a piece of silicone rubber with instant glue to a bacterial spreader (Fig. 4).
3. Use DNA low-binding tubes and low-retention pipet tips to increase sample recovery throughout the experiments. Protein low-binding tubes or siliconized tubes are also applicable for ChIP procedure before DNA purification.
4. To select histone modification to be analyzed, refer public ChIP-seq data in the ENCODE (https://www.encodeproject.org). To browse, click "Experiment" to open Experiment Matrix page, and sort data by "histone-ChIP-seq" in "Assay"-"Assay title" tab and by relating cell lines, tissues, or cell lineage of your interest in "Biosample" tab. To see

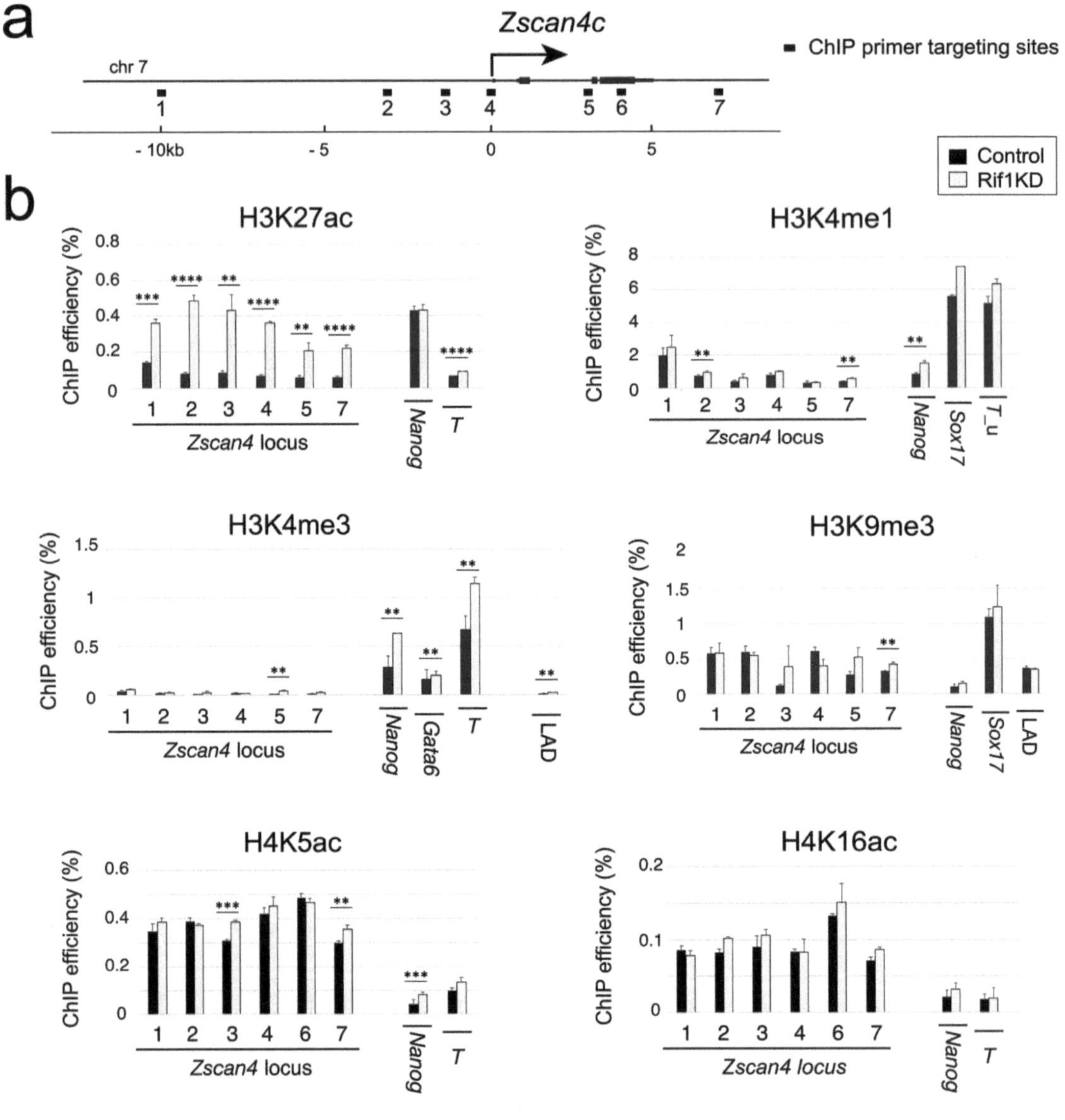

Fig. 3 Histone modification analysis of the *Zscan4* gene loci in Rif1-depleted cells. (**a**) A map of the representative *Zscan4* family gene (*Zscan4c*) and target regions of ChIP primers (top). (**b**) ChIP-qPCR analysis of histone H3 modifications at *Zscan4* target sites in Rif1-depleted ES cells. E14tg2a mouse ES cells were transfected with Rif1 siRNA or control siRNA twice and harvested at 72 h. The cells were subjected to ChIP-qPCR for histone H3 and H4 modification. In this figure, data of H3K27ac (active enhancer mark), H3K4me1 (primed enhancer mark), H3K4me3 (active promoter mark), and H3K9me3 (silencing mark) are shown. Black bars, control. White bars, Rif1 knockdown. Control target loci are as follows: *Nanog*, a pluripotent marker gene; *T*, *Sox17* and *Gata6*, differentiation regulatory genes; LAD, a LaminB-binding site; "*T*" targets the upstream promoter region, while "*T*_u" targets the enhancer region. Bars, SDs. The means of technical triplicates of representative data are shown and analyzed by using unpaired two-tailed Student's t-test. **, $P < 10^{-2}$. ***, $P < 10^{-3}$. ****, $P < 10^{-4}$. No marks, not significant. Bars, SDs. (Modified from [20])

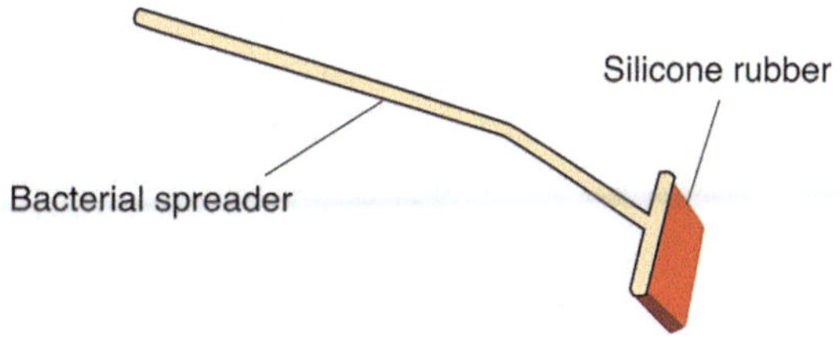

Fig. 4 Silicone rubber-attached scraper (*see* **Note 2**)

individual data, click "List" or "Report" to make experiment list and click experiment title or ID to show the experimental summary, links to information of used antibodies and cells, and sequencing reads data on genome browser. Previously reported modifications can be use as positive controls in your study. Well-examined six histone modifications are also subjected by International Human Epigenome Consortium (http://ihec-epigenomes.org/welcome/). It is important to validate antibodies in pilot experiments by using your own samples. Highly specific monoclonal antibodies (for example, used in [20]) also work well.

5. One μg of antibody is sufficient per one ChIP for abundant modifications, but optimize the amounts of antibodies before use. Calculation of antibodies and beads is as below. If the sample number is 2 (e.g., control and treatment) in 3 replicates, required amounts of ChIP antibody is 1 μg × 2 × 3 = 6 μg, and the volume of beads is 20 μL × 2 × 3 = 60 μL. Increase the amounts of antibody and beads by × 1.1 to prevent from shortage caused by aliquoting. Up to 600 μL beads per 1 mL binding buffer can be used for antibody binding in 1.5 mL tube. For analysis of rare histone modification, increase the antibody (to 2 μg per 20 μL of beads, or approximately 50% of capacity of beads), the volume of antibody-bound beads, and/or the input chromatin per ChIP.
6. Prepare antibody-bound magnetic beads just before the experiments. This step can be carried out during the preclear of ChIP Input in Subheading 3.3, **step 8**.
7. Most commercially available reagents are applicable, but check the sensitivity and amplification efficiency by using your ChIP Input DNA as a template in pilot experiments. Follow the instruction of reagents and real-time PCR system.
8. If the gene cluster harbors many and/or long TEs or repetitive sequences, retrieve several-fold extra length to search gene-surrounding regions. How long extra sequence to get can be decided after you identify the conserved regions (*see* Subheading 3.1.2). Masking repeats by colors or lower case upon downloading helps your primer design.

9. It is preferable that the surrounding region of the prototype gene is also conserved and in simple genomic structure without many TEs and repeats. If not, flanking regions of another family gene can be used as prototype sequences for primer design in upstream and downstream loci.
10. The examples of general rules of primer design are as follows: primer size, 17–25 mers; amplicon size, 80–150 bp; and GC contents, 40–60%. Avoid >3 bp homology sequence in a primer or in pairs. G/C at 3′-end is preferable but not T. Check primers not to target other genomic loci by using BLAST (https://blast.ncbi.nlm.nih.gov/Blast.cgi).
11. Due to sequence variation or high multiplicity of family genes, it may be difficult to assign common primer sets at the target sites to cover all family members. At minimum, sequence with more than 90% identity (and no mismatch in 3′ end) in most of family gene members can be considered as putative common primers.
12. By using gene-specific primers, histone modification at certain family genes can be compared to other loci. For example, specific primers targeting a family gene which has unique insertion of endogenous retroviruses nearby can distinguish histone modification levels from that of other family gene loci. This may help us evaluate the local effects of specific TEs on chromatin states.
13. If ChIP-qPCR seems to be not specifically amplified (e.g., the melting peak is multimodal or single but with minor shoulders), and the alternative target site is difficult to pick up, try to optimize PCR condition to increase the primer specificity, increase annealing temperature, or change program to three-step PCR instead of two-step. Check the amplification efficiency is near to 2.0 in optimized conditions.
14. In general, 1×10^6 cells are required per one ChIP assay for pan histone H3, pan H4, and abundant histone modifications, such as H3K4me3, K3K9ac, H3K27ac, H3K9me3, H3K27me3, or H4K20me2 for mouse ES cells. Several times more numbers of cells are required for rare modifications, such as simultaneous acetylation of H4K5/K8/K12. Since the cell types, treatment, and antibodies affect the detection limits, optimize the experimental scales and conditions by using your antibodies.
15. Sedimentation of chromatin fraction after nuclear lysis requires higher centrifugal force for other cell lines. For ES cells derived from C57Bl6/j mice, we centrifuge cells at $1750 \times g$ for 10 min.
16. Shearing chromatin is a critical step of ChIP experiment. Different types of sonicators have distinctive features and

performance, and chromatin shearing efficiency depends on cell types. Optimize the sonication condition on your apparatus by using the same cell lines at the same scale, identically processed as in Subheading 3.3, **steps 1–6**. Two sonicator combination protocol shown here is beneficial to shear chromatin appropriately in relatively short time. We usually process samples at the density of approximately 2–5 × 10^7 cells/mL. For a large number of input samples, water bath-type sonicator may be useful.

17. Check the size of sheared chromatin as follows: (1) Take small portions (e.g., one hundredth to thousandth of total sonicates) and add elution buffer to fill up to 27 μL. (2) De-crosslink at 65 °C for 3 h. (3) Add 50 μg/mL of RNase A and incubate at 37 °C for 15 min. (4) Add Proteinase K at 1 mg/mL and incubate at 56 °C for 30 min. (5) Take 5 μL out of total 30 μL digests and add 5 μL of 2× glycerol loading buffer. Load DNA onto 1.5% agarose gel in TBE with a 100 bp ladder DNA size marker and resolve them. (6) Stain DNA with DNA intercalater (ethidium bromide or other reagents).
18. For longer storage, samples should be stored as propanol (PrOH) precipitates. If ChIP DNA was sufficiently recovered, phenol-extracted samples can be divided and separately precipitated with PrOH for later reanalysis of other target loci, and stored at −80 °C.
19. ChIP DNA in aqueous solution is unstable and should be used up in a few days.
20. For ChIP-qPCR, standard curves should be made by using Input DNA for all primer sets to quantify ChIP DNA. Ideally, standard curves cover the Threshold Cycle (Ct) or Crossing Point (Cp) values from 20 to 30 by four dilution series. If Ct/Cp of standard samples is less than 20, add one or more dilution. In main experiments, at least three diluted Input samples are required for standard curves.
21. Required numbers of reaction are [Input DNA series (3 or 4) + ChIP samples to test (n)] × [technical replicates] × [factors for extra volume of premix (× 1.2)]. For 384-well plate, the formula is as follows: Mix 0.12 μL of primer set stock solution, 3 μL of 2 × enzyme-reagent mix, and fill up with water to 5 μL per reaction. Make premix for required numbers of reaction, dispense 5 μL of premix in 384-well plate, and add 1 μL of ChIP DNA (original or diluted) per well.
22. ChIP efficiency and fold enrichment: In general, ChIP efficiency (ChIP DNA per Input DNA ratio [%]) is the standard indicator. Fold enrichment is the ratio of values against negative samples, negative regions of the same histone marks, or the same regions for nonspecific antibodies. Since the fold

enrichment ratios often widely vary due to unreliable low values of negative samples, only use them in appropriate situation. In addition, modified histone marks versus pan histone marks are often shown. Generally, there is little difference in normalized and non-normalized data unless the histone deposition decreases by treatment. However, we recommend that one analyze pan histone levels in each experiment especially if you target the low-mappability regions, where public pan histone ChIP-seq data is usually not available.

Acknowledgments

We gratefully acknowledge Hideya Kawaji (Tokyo Metropolitan Institute of Medical Science) for critical reading of the manuscripts. This work was supported by JSPS KAKENHI grant nos. 17H01418 and 26251004 (H. M.), JSPS and SNSF under the Joint Research Projects (JRPs) (grant number JPJSJRP20191508) (H. M.) and JSPS KAKENHI grant nos. 15 K07164 and 23570243 (N.Y.-S.).

References

1. Kouzarides T (2007) Chromatin modifications and their function. Cell 128(4):693–705. https://doi.org/10.1016/j.cell.2007.02.005
2. Hecht A, Strahl-Bolsinger S, Grunstein M (1996) Spreading of transcriptional repressor SIR3 from telomeric heterochromatin. Nature 383(6595):92–96. https://doi.org/10.1038/383092a0
3. Schübeler D, MacAlpine DM, Scalzo D, Wirbelauer C, Kooperberg C, van Leeuwen F, Gottschling DE, O'Neill LP, Turner BM, Delrow J, Bell SP, Groudine M (2004) The histone modification pattern of active genes revealed through genome-wide chromatin analysis of a higher eukaryote. Genes Dev 18(11):1263–1271. https://doi.org/10.1101/gad.1198204
4. Santos-Rosa H, Schneider R, Bannister AJ, Sherriff J, Bernstein BE, Emre NC, Schreiber SL, Mellor J, Kouzarides T (2002) Active genes are tri-methylated at K4 of histone H3. Nature 419(6905):407–411. https://doi.org/10.1038/nature01080
5. Weinert BT, Narita T, Satpathy S, Srinivasan B, Hansen BK, Schölz C, Hamilton WB, Zucconi BE, Wang WW, Liu WR, Brickman JM, Kesicki EA, Lai A, Bromberg KD, Cole PA, Choudhary C (2018) Time-resolved analysis reveals rapid dynamics and broad scope of the CBP/p300 acetylome. Cell 174(1):231–244.e212. https://doi.org/10.1016/j.cell.2018.04.033
6. Creyghton MP, Cheng AW, Welstead GG, Kooistra T, Carey BW, Steine EJ, Hanna J, Lodato MA, Frampton GM, Sharp PA, Boyer LA, Young RA, Jaenisch R (2010) Histone H3K27ac separates active from poised enhancers and predicts developmental state. Proc Natl Acad Sci U S A 107(50):21931–21936. https://doi.org/10.1073/pnas.1016071107
7. Probst AV, Almouzni G (2011) Heterochromatin establishment in the context of genome-wide epigenetic reprogramming. Trends Genet 27(5):177–185. https://doi.org/10.1016/j.tig.2011.02.002
8. Bernstein BE, Mikkelsen TS, Xie X, Kamal M, Huebert DJ, Cuff J, Fry B, Meissner A, Wernig M, Plath K, Jaenisch R, Wagschal A, Feil R, Schreiber SL, Lander ES (2006) A bivalent chromatin structure marks key developmental genes in embryonic stem cells. Cell 125(2):315–326. https://doi.org/10.1016/j.cell.2006.02.041
9. Liu X, Wang C, Liu W, Li J, Li C, Kou X, Chen J, Zhao Y, Gao H, Wang H, Zhang Y, Gao Y, Gao S (2016) Distinct features of H3K4me3 and H3K27me3 chromatin domains in pre-implantation embryos. Nature 537(7621):558–562. https://doi.org/10.1038/nature19362

10. Kimura H (2013) Histone modifications for human epigenome analysis. J Hum Genet 58(7):439–445. https://doi.org/10.1038/jhg.2013.66
11. Black JC, Van Rechem C, Whetstine JR (2012) Histone lysine methylation dynamics: establishment, regulation, and biological impact. Mol Cell 48(4):491–507. https://doi.org/10.1016/j.molcel.2012.11.006
12. Park PJ (2009) ChIP-seq: advantages and challenges of a maturing technology. Nat Rev Genet 10(10):669–680. https://doi.org/10.1038/nrg2641
13. International Human Genome Sequencing Consortium (2001) Initial sequencing and analysis of the human genome. Nature 409(6822):860–921. https://doi.org/10.1038/35057062
14. McCarthy EM, McDonald JF (2004) Long terminal repeat retrotransposons of Mus musculus. Genome Biol 5(3):R14. https://doi.org/10.1186/gb-2004-5-3-r14
15. Teissandier A, Servant N, Barillot E, Bourc'his D (2019) Tools and best practices for retrotransposon analysis using high-throughput sequencing data. Mob DNA 10:52. https://doi.org/10.1186/s13100-019-0192-1
16. Lerat E, Fablet M, Modolo L, Lopez-Maestre-H, Vieira C (2017) TEtools facilitates big data expression analysis of transposable elements and reveals an antagonism between their activity and that of piRNA genes. Nucleic Acids Res 45(4):e17. https://doi.org/10.1093/nar/gkw953
17. Criscione SW, Zhang Y, Thompson W, Sedivy JM, Neretti N (2014) Transcriptional landscape of repetitive elements in normal and cancer human cells. BMC Genomics 15:583. https://doi.org/10.1186/1471-2164-15-583
18. Zalzman M, Falco G, Sharova LV, Nishiyama A, Thomas M, Lee SL, Stagg CA, Hoang HG, Yang HT, Indig FE, Wersto RP, Ko MS (2010) Zscan4 regulates telomere elongation and genomic stability in ES cells. Nature 464(7290):858–863. https://doi.org/10.1038/nature08882
19. Falco G, Lee SL, Stanghellini I, Bassey UC, Hamatani T, Ko MS (2007) Zscan4: a novel gene expressed exclusively in late 2-cell embryos and embryonic stem cells. Dev Biol 307(2):539–550. https://doi.org/10.1016/j.ydbio.2007.05.003
20. Yoshizawa-Sugata N, Yamazaki S, Mita-Yoshida K, Ono T, Nishito Y, Masai H (2021) Loss of full-length DNA replication regulator Rif1 in two-cell embryos is associated with zygotic transcriptional activation. J Biol Chem 297(6):101367
21. Sambrook JF, E.F., Maniatis, T. (1989) Molecular cloning: a laboratory manual, 2nd edn. Cold Spring Harbor Laboratory Press
22. Dahl JA, Collas P (2008) MicroChIP--a rapid micro chromatin immunoprecipitation assay for small cell samples and biopsies. Nucleic Acids Res 36(3):e15. https://doi.org/10.1093/nar/gkm1158
23. Langmead B, Trapnell C, Pop M, Salzberg SL (2009) Ultrafast and memory-efficient alignment of short DNA sequences to the human genome. Genome Biol 10(3):R25. https://doi.org/10.1186/gb-2009-10-3-r25
24. Langmead B, Salzberg SL (2012) Fast gapped-read alignment with bowtie 2. Nat Methods 9(4):357–359. https://doi.org/10.1038/nmeth.1923
25. The UCSC Genome Browser (2000) The UCSC Genomics Institute, the University of California Santa Cruze. https://genome.ucsc.edu/. Accessed 8 Oct 2015

Index

0-9, and Symbols

% Paraformaldehyde ... 88
(TTAGGG)n ... 106
.1 mM non-essential amino acids ... 165
.2% Triton X-100 ... 141
0.001% (w/v) pepsin 50 mM HCl ... 87
0.002% (w/v) pepsin solution ... 113
0.002% (w/v) pepsin solution in 50 mM HCl ... 106
0.05% digitonin ... 143
0.075 M KCl ... 2
0.1 M Na_3VO_4 ... 166
0.1 mM 2-mercaptoethanol ... 165
0.1% ProteaseMAX solution ... 157
0.2% Triton X-100 ... 57, 88
0.2% TritonX-100 in PBS ... 86
0.5 M EDTA (pH 8.0) ... 166
0.5 mM spermidine ... 143
0.8 M NaCl ... 141
1 × PBS (pH 6.8) ... 143
1 M DTT ... 166
1 M HEPES-KOH (pH 7.5) ... 166
1 M potassium cacodylate ... 57
1 M Tris-HCl (pH 7.5) ... 167
1 M Tris-HCl (pH 8.0) ... 166
1 mM $CaCl_2$ ... 141
1 mM EDTA ... 141
1 mM phenylmethylsulfonyl fluoride ... 154
1 mM sodium pyruvate ... 165
1 N HCl ... 166
1 μg/ml DAPI ... 107
1% formaldehyde ... 167
1% low melting temperature (LMT) agarose ... 67
1% ProteaseMAX solution ... 155
1% standard agarose gel ... 67
1.5% agarose gel in TBE ... 170
1× PBS ... 167
1× TBE ... 170
1× Tween Washing Buffer ... 129
10 mg/ml cytochalasin B ... 85
10% glycerol ... 141
10% neutralized PEI (Polyethylenimine) ... 144
10% sodium deoxycholate ... 166
10× phosphate-buffered saline (PBS) ... 166
100% ethanol ... 143
1000 U/mL of leukemia inhibitor factor ... 165
2 M glycine ... 167
2.5 M Glycine ... 129
2× Binding Buffer ... 129
2× glycerol loading buffer ... 167
20 mg/mL glycogen solution ... 170
20 mM HEPES-KOH (pH 7.2) ... 141
20% Nonidet P-40 ... 166
20% SDS ... 166
20% Triton X-100 ... 166
20% Tween-20 ... 166
20x SSC ... 76
2N HCl ... 76
2-propanol ... 170
2x SSC ... 76
3 M sodium acetate ... 170
3×Flag-pA-Tn5-Fl ... 144
37% Formaldehyde ... 129
3D chromatin structure ... 128
4 M LiCl ... 166
4 mM glutamine ... 165
4% Paraformaldehyde ... 112
5 M NAC-Ac ... 154
5 M NaCl ... 166
5 N KOH ... 166
5 N NaOH ... 166
5× Tris-borate buffer (TBE) ... 167
50 mM aqueous ammonium bicarbonate ... 157
50 μg/ml propidium iodide (PI) solution ... 67
53BP1 ... 94
6 N HCl ... 166
70% formamide solution in 2xSSC ... 113
75 mM hypotonic potassium chloride (KCl) solution ... 56
75% ethanol ... 170

A

Abnormal chromosome morphology ... 9
Accutase ... 155
Acentric chromosome fragments ... 84
Acentric fragments ... 11
Acridine orange staining ... 84

Eisuke Gotoh (ed.), *Chromosome Analysis: Methods and Protocols*, Methods in Molecular Biology, vol. 2519, https://doi.org/10.1007/978-1-0716-2433-3,

Actin inhibitor 83
Active gene loci 162
Active replicating DNA 117
Active replicating regions 117
Adapter sequences 141
Adenosine triphosphate (ATP) 48
Adherent cells 47
Agarose electrophoresis 65
Agencourt AMPure XP beads 143
Aging 54
Alcohol and acetic acid fixation 10
Alexa 594 conjugated goat anti-mouse IgG antibody 19
Alkaline comet assay 65–71
Alkaline running buffer 67
AlphaMEM 12
Alpha satellite centromere 86
Alzheimer's disease 153
Ampicillin (100 mg/mL) 144
AMPure XP beads 129
Anaphase bridge 11
Annexin 54
Annexin PI Method 57–58
Annexin V 54
Annexin V APC and annexin-binding buffer 56
Antibody-bound magnetic beads 170
Antibody buffer 143
Anti-CENP-C antibody 29
Anti-gH2AX antibody 86, 89
Anti-mouse IgG (H&L) secondary antibody 143
Apoptosis (a-po-toe-sis) 53–62
Arginine residue 161
Asymmetrical chromatid exchanges 11
ATR 18, 94
Atypical enhancer-like structure 161

B

Banding pattern 117
Beads loading method 117
Beads on a string 117
Binding buffer 141
Binucleated cells 87
Biotin-conjugated concanavalin A, Type IV 143
Biotin-14-dATP 129
BLAST 172
Blast formation 3
Blebbing 54
Blocking buffer 29
Bloom syndrome 74
BRCA1 94
BRCA1 foci 94
BRCA2 94
Breakage-reunion hypothesis 10
Breakup of the chromatin 54
Budding 54
Buffy coat 3

C

Caenorhabditis elegans 54
Cantharidic acid 47
Cantharidin 47
Carnoy's fixative 2
Carnoy's fixative solution (methanol and acetic acid = 3 vol: 1 vol) 56
Caspase-activating protein 55
Catalyst-donor solution 154
CBMN assay 83
CBMN method 84
CCCTAA 112
CCD camera 20
Cell adhesive 29
Cell cycle regulation 161
Cell cycle synchronization 28
Cell death 53–55
Cell division 17
Cell shrinkage and pyknosis 54
Cellular proliferation rate 18
CENP-B box 86
CENP-B box sequence 112
CENP-B Box staining PNA probe 87
Centric ring 11
Centromere FISH probe 86
Centromere-specific probes 83
Centromeric chromatin 27
Chemical catalyst 153–158
Chemical-induced premature chromosome condensation 41–48
ChIP binding buffer 170
ChIP efficiency 181
ChIP elution buffer 170
ChIP wash buffer 1 (NaCl) 170
ChIP wash buffer 2 (LiCl) 170
ChIP-qPCR analysis 161
ChIP-sequencing analyses (ChIP-seq) 162
Chitin resin 144
Chloroform 143
Chloroform-Isoamyl alcohol 151
Chromatid exchanges 10
Chromatin-bound proteins 161
Chromatin/chromosome-mediated cellular events 165
Chromatin condensation 54
Chromatin fragments 54
Chromatin immunoprecipitation (ChIP) 161
Chromatin immunoprecipitation with sequencing (ChIP-seq) method 161–165, 177, 182
Chromatin interaction 127

Chromatin structure 27
Chromosomal aberrations (CAs) 9
Chromosomal type aberrations 10
Chromosome 1–6
Chromosome alignment 27
Chromosome breakage 84
Chromosome condensation 27
Chromosome conformation capture (3C) 127
Chromosome missegregations 84
Chromosome rearrangement 127
Chromosome segregation 27
Chromosome territories 117
Classification of chromosomal aberrations 9–14
CLASTALW 172
Cleavage under targets and release using nuclease (CUT&RUN) 141
Click reaction staining for EdU 80
Cloudy nebula 117
Colchicine 18
Combinatorial organization in the functional features of chromatin 161
Comet assay 65–71
Comet Assay IV 69
Comet-FISH methods 69
Compartment shifts 128
Complete Protease Inhibitor Cocktail 143
Complex-type translocation 100
Comprehensive ChIP-qPCR analysis 164
Con A-conjugated Dynabeads 141–152
Con A-conjugated magnetic bead 141
Concanavalin A (ConA) 6
Control mouse IgG 143
Copy number variation 128
Counterstaining 29
CREST serum 84
CUT&Tag 141–152
CUT&Tag procedure 141
Cy3-dUTP 119
Cy3-dUTP-labeled S-phase PCCs 117
Cy5 118
Cytocentrifugation 33
Cytocentrifuge 28
Cytocentrifuge system 93
Cytochalasin B 83
Cytofunnel 33
Cytoglass 33
Cytokinesis block 83–90
Cytokinesis block micronucleus (CBMN) 84
Cytokinesis failure 11
Cytospin 33

D

DAPI 19
Daughter cells 27
dCTP 129
Denaturing solution 70% formamide in 2x SSC 76
Dense nuclear peripheral aggregations 54
Depolymerizes 36
Development 54
dGTP 129
Dialysis buffer 141
Dicentric 11
Differential staining method 73
Dig-300 buffer 143
Digitonin 143
Dig-wash buffer 143
Dissolved in dimethyl sulfoxide (DMSO) 29
DNA damage 65–71
DNA damage foci 93–97
DNA damage responses and repair 161
DNA damage signal probes 83
DNA damaged site 93
DNA double-strand breaks (DSBs) 10
DNA fold 117
DNA Polymerase I, Large (Klenow) Fragment 129
DNA purification column kit 170
DNA repair proteins 93
DNA replication 161
DNA transposons 163
Double thymidine block and release 36
dTTP 129
Dulbecco's Modified Essential Medium (DMEM) 12, 29
dUTP 56
Dynabeads MyOne Streptavidin T1 beads 129

E

EdU staining solution 76
Elimination of the cells 54
Elution buffer 141
Endogenous retrovirus MERVL 165
Endonuclease cleavage products 56
Endothal 47
Energy-dependent cascade 55
Ensembl 165
Environmental science 41
Eosin 54
Epigenetic information 161
Epigenome 141
Epigenome research 141
Epigenomic aberration 128
Escherichia coli strain BL21-pLysS Singles™ Competent Cells 144
Exchange hypothesis 10
Exponentially growing cell 93

F

FANCA ... 94
FANCB ... 94
FANCC ... 94
FANCD2 ... 94
FANCD2 foci ... 94
FANCE ... 94
FANCF ... 94
FANCG ... 94
Fanconi anemia patient cells ... 74
Fanconi anemia protein ... 94
Fetal calf serum (FCS) ... 2
Fibronectin ... 29
Ficoll-Paque method ... 5
Filling off ... 54
Flow cytometer ... 20, 155
Fluorescence Plus Giemsa (FPG) staining ... 13, 78
Fluorescent immunocytochemistry technique ... 93
Formamidopyrimidine DNA glycosylase enzyme (FPG) ... 66
Fragmentation of the nucleus ... 54
Fusion-PCC ... 42
Fusogenic virus ... 42

G

G1-PCCs ... 118
G2-PCCs ... 118
Gamma-H2AX ... 84
Gene cluster ... 161
Gene expression ... 161
Gene transcription ... 141
Genome architecture ... 127
Genome-wide chromatin structural aberration ... 127–138
Genome-wide chromatin structure ... 127
Genomic aberration ... 127
Genomic instability ... 106
Genotoxicity ... 83
GFP proteins ... 25
Giemsa dye solution ... 2
Glass Beads ... 119
Glu-C solution ... 155
GM05389 ... 119
Goat serum in PBS solution ... 19
Green fluorescence protein (GFP) ... 123
Gurr buffer ... 12

H

H3K14ac ... 162
H3K27ac ... 162
H3K4me1 ... 162
H3K4me3 ... 162
H3K9ac ... 162
H4K16ac ... 162
H4K8ac ... 162
Hank's Balanced Salt Solution (HBSS) ... 2
Heat inactivation ... 12
HEGX buffer ... 141
HeLa cells ... 36
Hematoxylin ... 54
HER2 ... 100
Heredity genetics ... 41
Heyflick's limit ... 105
Hi-C analysis ... 127–138
Hi-C Lysis buffer ... 129
Hierarchical chromatin structures
  domain ... 127
  loop ... 127
Highly homologous segmental multiplications ... 161
High Sensitivity DNA Chips ... 143
High-throughput sequencing analysis (ChIP-seq) ... 161
Histone ... 161
Histone acetylation ... 161
Histone H3 ... 18, 161
Histone H4 ... 161
Histone methylation ... 161
Histone modification analysis ... 161–182
Histone modification ChIP-assay ... 161
Histone modifications ... 141
Histone-modifying enzymes ... 153
  erasers ... 153
  writers ... 153
HOECHST 33342 ... 29
Homeostatic mechanism ... 54
Homologous recombination ... 74
Homologous segments ... 162
Homologous sequence ... 163
Homotetramer ... 149
Hypotonic buffer ... 28
Hypotonic lysis buffer ... 154
Hypotonic pretreatment ... 10

I

Immunostaining ... 28
Imstar Pathfinder ... 69
In situ Hi-C ... 128
Inaccurate mapping ... 161
Inactivated X chromosome ... 162
Inter-chromosome fusion ... 106
International Human Epigenome Consortium ... 179
Interphase chromatin ... 42
Intra-chromosomal chromatid fusion ... 106
IPTG (250 mM) ... 144

K

KAPA HiFi HS ReadyMix 129
KAPA HyperPrep Kits 129
Karyorrhexis 54
KCM solution 95
Kinetochore 27
Kinetochore formation 27
Komet software 69

L

Laser scanning microscopy 121
Lectins 6
Loss of homologous recombination activity 74
Low mappability 161
Low-mappability regions 161–182
Lysine residue 161
Lysis buffer 67

M

Macrophages 54
Maturation-promoting factor (MPF) 48
mBAND 100
MDC1 94
Membrane integrity 54
MG132 .... 29, 37
Microcystin-LR 47
Micronuclei 83
Micronuclei aberration analysis 83–90
Micronuclei formation 11
Micronucleus test 83
Microtubules 27
Milli-Q water 43
Minimum Essential Media (MEM) 12
Misrejoining 9
Mitotic donor cells 42
Mitotic index (MI) 17–25
Mosaic end-adapter A oligonucleotide (Tn5ME-A) 146
Mosaic end-adapter B oligo (Tn5ME-B) 141
Mosaic end-reverse oligonucleotide (Tn5MErev) 141
Mounting buffer 29
Mouse ES cells 162
Mouse monoclonal anti-BrdU antibody 57
MRE11 94
MRN foci 94
Multicolor FISH (mFISH) 100
Mutagenic studies 41
Mutagen screening 83

N

NaOH, 1mM EDTA pH 13.0. 67
NBS1 94
NEBNext High-Fidelity 2× PCR Master mix 143
Necrotic cell death 54
Nematode 54
Neutral comet assay 93
Neutralization buffer: 0.4 M Tris-HCl pH 7.4. 67
Next-generation sequencers (NGS) 127
Nocodazole 24, 29
Nonhomologous end joining (NHEJ) 74
Noninvasive manner 1
N,S-diacetylcysteamine (NAC-Ac) 153
Nuclear envelope breakdown 27
Nuclear lysis buffer 167
Nuclear lysis+ buffer 167
Nuclear segregation 83
Nucleosome core histones 161
Nylon mesh-bottomed device 154

O

OECD guideline 66
Oncology 41
Orphan chromosome 84
Overamplification 161

P

Paclitaxel 24
Pan-centromeric FISH DNA 87
Parafilm 30
Paraformaldehyde 19
PARP-deficient cells 74
Pathognomonic sign 54
pATM 94
PBS/0.1% Triton X-100 (PBS-Tx) 170
PBS/0.02% Tween-20 (PBS-Tw) 170
p34cdc2/cyclin B complex 48
p53-dependent pathway 54
PEG-LANA-DSSMe 153
Peptide nucleic acid (PNA) 100
Perforin-granzyme-dependent killing 55
Peripheral blood lymphocytes (PBLs) 2
Permeabilization buffer 28
Phagocytosis 54
Phagolysosomes 55
PHA-M 5
Phenol/chloroform/isoamyl alcohol 170
pHH3 immunocytochemistry 21
Phosphate buffer 2
Phosphate-buffered saline (PBS) 2
Phospho-NBS 94
Phosphorylated histone H3-S10 (pHH3) 18
Physiological death 53
Phytohemagglutinin-P (PHA-P) 2, 5
PNA probe 106

PNA probe solution 87
Polycomb 162
Polyethylene glycol (PEG) 42
Poor suspendability 141
Post transcriptional modification 27
Posttranslational modifications 161
Posttranslational modifications (PTMs) of histones 153
POT1 111
Premature chromosome condensation 41–48
Prematurely condensed chromosomes (PCCs) 44, 117
Programed cell death 54
Prometaphase arrest 36
Propidium iodide 54
Propionic anhydride solution 155
Protease inhibitor 154
Proteinase K 129, 143, 170
Protein-A Tn5 transposase fusion protein 141
Protein semisynthesis 153

Q

Quantitatively measure the DNA breakage 65
Quantitative real-time PCR (qPCR) 170
Qubit dsDNA HS assay kit 143

R

Rabbit polyclonal anti-phospho-histone H3 (Ser10) antibody 19
Rad50 94
Rad50 foci formation 94
Rad51 94
Rad51C 94
Rad51D 94
Rad51 family genes 94
Radiation biology 41
RAP1 111
Real-time PCR 128
Recipient interphase cells 42
Reciprocal exchanges 12
Reciprocal translocation 12
Relaxation of supercoils 65
Repeat-enriched gene cluster regions 161
Repetitive sequences 161
Replication foci 117
Restriction enzyme MboI 129
Retroviruses 163
Reverse-crosslinking buffer 129
Rifl 161
RNase A 143, 170
Roche complete protease inhibitor cocktail 129
Roswell Park Memorial Institute (RPMI) 12

S

Sample-unique primer sets 143
Scanning confocal microscope 117
SCE frequency 74
Scleroderma patients 84
Sendai virus 42
Shelterin proteins 111
Signaling pathway 27
Single-stranded breaks 65
Sister chromatid 73
Sister chromatid exchange (SCE) 73–81
SlowFade 19
Sodium dodecyl sulfate (SDS) 129
S-PCCs 118
S-phase chromosome 117–124
Spindle 27
Strand-specific telomere FISH 106
Striped arrays of banding structured chromosome 117
Structural aberrations 127
Submegabase-scale chromatin domain structure 127
SUMOylation 161
Suspension-type cells 47
SYBR 88
Synchronization of cell cycle 36–37
Synthetic histone acetylation reaction 154

T

TAD boundary disruptions 128
Tagmentation buffer 143
Tail moment 66
Taxol 29
T-cell lymphocytes 3
T-cell-mediated cytotoxicity 55
T4 DNA Ligase 129
Telomere 111–115
Telomere breaks 106
Telomere exchange 106
Telomere fusion 106
Telomere-specific probe FISH 106
TERF2 111
Terminal deoxynuleotidyl transderase (TdT) 57
Terminal dUTP nick-end labeling (TUNEL) 56
Tertiary and quaternary structure of the DNA 65
Tetramethylrhodamine 155
Thymidine 29
Thymidine analog 73
Thymidine block 36
Time lapse recording 20
TIN2 111

T-loop 105, 111
TNE buffer 170
Topologically associating domains (TADs) 127
TPP1 111
Transcription factors (TFs) 141
Translocation 100, 128
Transposable elements (TEs) 162, 163, 167
TREF1 111
Tris-HCl (pH 8.0) 143
Tris-HCl buffer (TB) 170
Tritiated thymidine (^{3}H-TdR) 73
Triton X-100 19
Trypan blue 54
Trypsin-EDTA 19
Trypsin gold solution 155
Trypsinization 44
TTAGGG region 111
Tumor necrosis factor (TNF) 55
Two-cell stage of mouse zygote 164
Two-color FISH 100
Type 2 diabetes 153

U

Ubiquitination 153, 161
UCSC Genome Browser 165
UltraPure Phenol-Chloroform-Isoamyl alcohol 143
Unrepaired DNA damages 9

V

Vacutainer® CPT™ centrifuge tube 2
Venipuncture 1

W

Wash buffer 143
Watson-Click semiconservative DNA replication model 73
Whole chromosome painting 99–103

X

X-lysis buffer 167
X-lysis+ buffer 167
X-lysis+ buffer minus DTT 170
XRCC1 74
XRCC2 94
XRCC3 94

Z

Zscan4 161
Zscan4 family gene loci 161

MIX
Papier aus verantwortungsvollen Quellen
Paper from responsible sources
FSC® C105338

If you have any concerns about our products,
you can contact us on
ProductSafety@springernature.com

In case Publisher is established outside the EU,
the EU authorized representative is:
Springer Nature Customer Service Center GmbH
Europaplatz 3, 69115 Heidelberg, Germany

Printed by Libri Plureos GmbH
in Hamburg, Germany